# MAN DESIGNED HIMSELF

**By Paul Tatham**

Published by Lulu.com

First published June 2012.
This edition March 2017


© Paul Tatham 2012.

ISBN  978-1-4452-5263-6

# CONTENTS

About the author.
Paul Tatham is an engineering graduate of Imperial College, London University
and is now retired after a career in Xerox Corporation and the computer industry.

# INTRODUCTION

This book came about because of my wish to know how things work. It is my nature to solve things for myself, rather than learn from existing technical books, and because of my basic process of starting with a blank sheet of paper and doing a complete review of everything, I seem to have been able to identify scientific errors

In my first book – **'A New Theory of the Universe'** - I showed that Einstein's theories of Special and General Relativity are incorrect, his Time Dilation is incorrect and space is not warped and so is not the source of gravity. Mass is also not understood as it is not the cause of resistance to motion. These misunderstandings have existed for nearly one hundred years, and are the reason scientists to fail to grasp the cause of Gravity, Dark Matter and Dark Energy.

In this book I show that scientists may have another misunderstanding and it is that electrons in sodium ions passing through a nerve must carry a code of energy levels in their field for identifying pain location and copying DNA.

There is a match of energy, and a single common strand of DNA in the genes, between a cell and its neighbours, that enables molecules to hold cells together. When this energy is exhausted and there is a risk of cells falling apart, cell voltage reduces so that sodium ions enter the cell with more energy. Thus cell division occurs when the energy stored in a cell's DNA is exhausted, and new energy is required for cell replacement. It is a process of energy flow that the ancient Chinese fully understood, and it is why acupuncture evolved.

These misunderstandings have led to the failure to fully understand the cell growth process and therefore how it may be possible to stop any cancer growth simply by dispersing the electrical energy in the nerve or tumour that the growth requires.

Also, I propose that the accepted theory of random gene mutations, where some are favourable and others not, leading to 'survival of the fittest', is not what led to evolution. Instead, there are at least four routine causes; embryonic cell multiplication, cell interdependence, cell damage and muscle use, and I discuss why gene mutation is a natural and structured process of adaptation that always produces favourable results. An animal species designs itself based on 'need'.

This natural process of gene adjustment is why animals have evolved so quickly and so precisely to perform their task of survival. The same is true for plants where changes in sunlight and damage caused by such things as hurricanes cause changes to DNA and rapid adaptation to the changed environment. Errors may well occur during gene copying, but I suggest that these are trivial in terms of evolution.

In recent years there has been much discussion about the alternative views of 'Creation', 'Intelligent Design' and 'Darwin's Natural Selection'. This book argues in favour of Darwin, but I suggest that "Natural Selection" is insufficient to explain the vast complexities of life, and if 'Intelligent Design' is nonsense, how exactly did the incredible devices such as the eye and the brain come into existence? I seek to show in this book how these things may have evolved and that it was the free choice of each animal that led to such rapid evolution, i.e. It was not 'Intelligent Design', it was 'Self Design'.

This is not a science textbook. There are no complex equations, chemical formulae or unpronounceable words and there are no processes discussed that are already well known and can be found in any textbook, such as photosynthesis and respiration. The book discusses only processes that might not be adequately understood. It is a set of theories based on the application of pure logic to known scientific facts.

Enormous progress has been made in understanding cancer in recent years but I have given it more thought simply because my wife died of cancer many years ago. I have reached some conclusions that I believe could slow growth and perhaps eventually shrink all cancer tumours, and prevent most cancers from starting. Cancer is all about energy. It is the result of the lifestyle that we now chose to live, and I discuss why this is the case.

At the end of the book I consider that if life evolved completely naturally and there is no God, why does religion flourish so strongly? Clearly the concept of 'God' satisfies man's emotional need and the need for hope, teaching and discipline, and it was therefore natural for such a concept to have reached a significant level of belief, but belief in some parts of the world has been a dominant part of culture for so many generations that perhaps it is now an 'instinct' embedded in the genes, and children are born with the belief firmly established in their mind, just as a bird knows it must migrate south in winter. Religion thwarts progress because 'contentment' is achieved without material progress, but no matter how hard the atheists try to convince believers that there is no God, the instinct factor will ensure that religion will exist for many more centuries.

Paul Tatham

## Chapter One

### HOW THIS BOOK CAME ABOUT.

My first book, **'A New Theory of the Universe'** was about space, time and gravity. This second book is about evolution. Both were written just to satisfy my own curiosity; to understand better how the world came to be as it is.

It was not my intention to write either book. In the case of my earlier book, I simply sat in my garden thinking logically about gravity, space and time and fairly quickly reached conclusions that were different from those of the scientific community. Then, having made some unexpected progress I began to look deeper into atoms and particles and discovered that a fundamental misunderstanding seems to have been made by scientists in resistance to motion (mass) and in Einstein's theory of General Relativity and time dilation.

These errors have caused scientists to search for solutions via complex mathematics and this has led to the theories of multiple universes, multiple histories and multiple dimensions, all of which I believe are mathematically-generated absurd nonsense! The universe is not so complicated and in my opinion, when the 'errors' are corrected, everything begins to make sense.

This second book was again written for my own purposes and came about for two reasons. Firstly my wife died of cancer several years ago and that was an awful waste of an intelligent and healthy life, yet although a massive amount of excellent work has been done and can be found on websites in this extremely complex subject, no one has yet been able to be precise about all of the causes, or what actually changes within a gene. So I began to think it through for myself but with no great ambition to write a book.

The second reason was more of a joke. When I saw the photos of my daughter's wedding in 2004 I said to myself "I must do something about my hair" because I haven't got much! I made a New Year Resolution in January 2005 to try to grow new hair. So both of these reasons involved finding out more about the process of growth.

When I was in the middle of writing my book about the universe and had already concluded what was the source of gravity. I sat out on my deck in the Bahamas one warm night in February 2005 and realised that if gravity was what I felt it was, the evolution of the nervous system and the creation of an eye become a little more understandable.

Biology and physics are closely linked and the evolution of life cannot really be understood until all the forces of the universe are properly understood. It was all the different energies within the universe that led to the evolution of life.

Biologists have known the importance of the sun in the process of photosynthesis in plant growth for decades, and also that humans need the sun to produce vitamin D and other chemicals that allow the brain to switch off after dark and wake up in daylight, but perhaps it is because physicists have failed to make the complete link between the various energies available to life, certain pieces of evolution remain a mystery?

So this book is an attempt to link the forces that I believe exist in the universe, as described in my first book, to the evolution of life. It is an attempt to fill in the missing blanks. As before, I am not interested in the detail of chemical formulae or mathematical equations, I am interested only in processes, i.e. it is a 'top down' examination of the energies and forces available to life forms that enabled evolution from the simple amoeba to intelligent man.

Note that I have used the life-form 'amoeba' throughout the book simply as a name for early pond life that most people will know. It is actually a bad choice because amoeba do not grow, they simply divide into two, but you know what I mean..

Also note that I understand physics far better than complex biochemistry and so I have tended to use equivalent physics terms as an analogy in my explanations of processes, which I hope biochemists will understand.

The human body has evolved around all sources of energy and if for some reason the levels of energy are too high – such as when you sit in the sun all day – the body's cells cannot cope and begin to go wrong. Thus the energies of both gravity and the sun may produce answers to some of the illnesses that we suffer yet cannot identify the cause.

To satisfy my objective of showing that evolution of an intelligent life form was a natural outcome it is not essential for me to be exactly correct on any scientific suggestion; rather it is simply necessary to show that there are entirely feasible solutions. Thus there will be many statements in this book with which scientists may not agree. That is bound to be the case because I do not have a degree in Biochemistry, nor have I studied sufficiently to validate my conclusions. However there are also some concepts of scientists that I do not entirely agree with, although I must add that I am impressed by their immense progress in a very complex subject and their findings to date.

## My background and approach.

I am retired and to make my retirement more comfortable I bought some land in the Bahamas in 2000 and had a house built, so now I spend every English winter in the sun. My background is that of an engineer with a degree from Imperial College, London University. I spent most of my working life with Xerox Corporation and the computer industry. That background did not give me any knowledge about either the universe or evolution but it did teach me how to think logically and solve problems and that is how I am now attempting to provide solutions to problems.

My process for writing the book was exactly the same as that for my first book. My technique was to start with a blank sheet of paper and write down everything that is definitely accepted as being factually correct. Then look at clues from life and begin to fill in some of the blanks. Then consider the process of evolution that must have occurred to produce man, and only when all the processes seem to be clear in my mind, read some textbooks to compare my results with those of scientists. And that is broadly how this book is laid out. My approach is to understand processes from their first principles rather than to read websites on what others have done.

It was the use of pure logic and the clarification of processes based on known facts rather than looking at the detail of chemical reactions and equations. Again it was a 'top down' process and examined the entire spectrum of life in order to gather clues that would allow logic to bring results.

My logic ran into dozens of pages but I will show a small sample here so that you can see the kind of process I followed and each section of the book will show further examples on how I reached a decision on a particular aspect.

## Some examples of my logic.

1. Darwin believed as I do that 'use' and 'non-use' was a key part of evolution. That means the use of a muscle, sense, or decision in the brain somehow feeds back to make a gene a greater priority in the next generation.
2. Cancer is known in some cases to be due to the possession of particular rogue genes, but these exist in every cell not just the area of the cancer, therefore some cancer genes can be created by the parent and passed on to the offspring.
3. To pass cancer genes on they must be manufactured or copied somewhere other than in each cell at the time of cell division and growth and that would have to be in the reproductive organs.
4. The automatic instinct of insects and the energy of 'use' signals to muscles suggest that energy must somehow go to the reproduction organs, and cancer genes are either produced in those organs or are produced when male and female genes are mixed and selected for the offspring.
5. But the rogue cancer genes seem to be extra to requirement because no organs or functions seem to be lost when they are produced, so maybe the mix of male and female genes either produces or allows rogue genes to be handed down without affecting essential genes, or there are many unused genes in the genome.
6. And the simplest form of cancer of the skin (the BCC) is clearly caused by

an excess of sunlight. Therefore it seems logical that all cancers are due to excess energy, or a level of energy that matches that required for a rogue cancer gene to take control of a cell.

7.  The sun provides energy to plants and we evolved using some of the same genes as plants therefore perhaps the sun plays a bigger role in our own evolution than we think. So have animals adjusted the plant processes to allow movement and relocation without risk to life?

8.  Then the question arises; is it the wavelength of the sun or the level of energy in each 'packet' of sunlight that influences things? Or is it just the total energy of all wavelengths?

9.  But as we are known to have thousands of genes and billions of neurons in the brain, it is not possible for the limited number of wavelengths of the sun to produce such a range of things. But it would be possible for the sun's wavelengths to have produced a range of different neurons, motor neurons and 'controlling' genes.

10. Genes are certainly not identical and neither are their proteins. If these are different then the energy required to make them must be different.

11. So energy is important, and it is not just the number of electrons in each atom that is important, it is the size of the field that each one carries, enabling the construction of a particular molecule.

Those few examples show the kind of logical thought process I went through and as you can see it is very much like doing Su Doku although perhaps a little more useful! It is only when all the clues are written down that you can begin to see some kind of solution emerging.

From all this analysis, and without reading any textbooks, I reached my own conclusions on how evolution occurs. This result was different to that reached by scientists and so further analysis was needed to define which is correct.

## My key findings and how they differ from scientists opinions.

It was at a late stage, after completing the first draft of this book, that I concluded that, if my solutions are right, there seems to have been a lack of the application of basic physics and chemistry by scientists in deciding the processes of how cells work, where they get their energy from, why the must renew themselves and how genes mutate. And having now found two main branches of science with the similar problems I wonder whether the 'Peer Review' process is adequate as this should reveal such errors? Perhaps the review should include a person from a completely unrelated discipline such as engineering?

One of the advantages of being an engineer is that we are naturally inclined to find out how things work and how to repair things that don't work. Also we have both feet firmly on the ground of reality! No flimsy theories or wacky ideas such as 'multiple universes' and 'curved space'!

The main error can be summarised as not realising that an electron passing through nerves must carry energy in its electromagnetic field, and that it is this wide variety of energy, not the electron itself, that nature has adopted and used to identify pain location and the production of specific proteins in cells by Ribosome and RNA. But I will discuss that further later.

At this time I will give some very brief examples of my own conclusions below,

- The eye is the source of sunlight entering the body to supply essential energy to recharge neurons and cells for growth.
- In animals, DNA in all genes is created to a specific level and amount of energy by neurons which originally collected their energy through the eyes. Now the neurons continually recharge themselves, monitor the sun's strength, muscle and sense use and cell repair and adjust DNA in the genes in sperm and egg to improve cells for the offspring.
- The purpose of the petals of flowers and leaves if there are no flowers, is the same as eyes, to collect the strongest wavelengths of the sun and pass this energy to the seed pods in order to adjust their DNA energy level to the optimum for the climate at their location.
- Cell division and growth will not be started by the controlling gene until the molecular energy bond between neighbouring cells begins to break down. The right wavelengths of energy in the right quantity are then provided by electrons in ions in the nerves, to restore the bond by producing new DNA.
- Neurons play a much bigger role in life than is currently understood. Some have their level of energy fixed at birth and some have a fixed energy that is inherited. Some are able to adjust their normal energy level within a narrow bandwidth.
- Gene mutation in animals is primarily as a result of changes in the strength of the sun in different location, the deliberate conversion of ordinary cells into muscles and the extent of their use, and of course viruses and carcinogens. It is not a random process.
- In plants, gene mutation is primarily caused by the sun and by what I call 'reverse engineering' where a plant has survived damage during the growth season but the damage has changed the genes in the leaves and the changed level of energy and mix of the DNA is passed to the seeds so that the new plant has a better chance of survival.
- The nervous system was created in two ways. The spinal nerves were

11

created moving in the Earth's gravity (or the Earth's magnetic field) causing electrons to move and convert ordinary cells into energy carriers. The peripheral nerves were started the same way but were extended by cells drawing their need of the sun's energy from other cells so that the pond-life could grow fat and so converting intermediate cells into nerves.

- Muscle use defines the shape of the entire body and so a muscle gene has control in practically every cell surrounding the muscle. It is conversion of pain cells into muscles, and then muscle use driven by the need for survival, that creates new animal species.

When these things are understood, Cancer, Parkinson's disease, re-growing spinal nerves and Astrology may become a little more understandable.

## Choice produces change.

I can summarise a little at this point by saying that every animal has a choice in what food to eat and how to protect itself from predators. The choice may be limited by circumstances and may be only a single choice to allow survival, but nevertheless it is a choice.

In my opinion, once the choice has been made, evolution will take over and give the animal all the physical strength it needs to maximise the benefit of that choice. The choice I describe is not one that requires a brain or positive thinking, nor is it an event of random mutation, it is simply a change in direction brought about by circumstances.

Plants do not have this choice because they cannot move and it is a case of grow or don't grow. So the choice is available only to animals and we humans are still making such choices today.

For example; we can walk or take the car. We can make things or buy things. We can do research and find out for ourselves or we can wait to be told. We can be a sporting person or a bookworm. We can lie in the sun all day or sit in the shade wearing sunglasses.

In many cases inherited instinct will guide our choice and so we excel in what our parents did, but we still have the choice.

Each of these typical choices will drive evolution towards a different design of person. The car traveller will produce children with short weak legs and a fat body. The bookworm children will have a big head and weak limbs whereas the sporting person's children will have big muscles and a smaller head. (I hope that does not seem like an insult)! If we make things rather than buy them our brains develop and so our heads will become bigger.

The common theme to all of these is 'use'. Darwin proposed that 'use' and 'non-use' were drivers of evolution but unfortunately the science community did not accept his ideas because no link could be found. I hope to correct that in this book.

One interesting conclusion I reached is that every animal is completely content with what it is! A cow is completely happy being a cow. If it needed to have additional functions for survival it would eventually make the adjustments, although it would take a long time. So the choices are driven by circumstances or the need to survive, not by some kind of intelligent wish by the animal to be something else. One could say evolution is driven by fear. And I would not describe it as 'survival of the fittest', rather it is 'survival of those who try hardest' and so evolve fastest.

Any grandparent can decide to produce the world's best tennis player or the world's fastest runner and provided that each generation pursued the same goal by seeking to achieve the goal themselves through training and muscle development, the grandchild would be close to achieving the goal. It's just a matter of prioritising the genes.

The next chapter suggests that there are clues that help us to understand the evolution of animals. These clues are then developed into theories which I discuss in later chapters.

You will note that I switch constantly between animals and plants during the book. The reason is that both must use similar processes in evolution and making such comparisons helps to define the processes as well as the 'process improvement' that animals have developed in order to evolve faster.

# Chapter Two

## THE KEY PRINCIPLES.

My first book on the universe taught me that scientists can make fundamental mistakes. If I were simply to study the scientist's opinion on evolution then I would be adding nothing to progress science because in two years I would end up at the same position and opinion as they are now, and as I am not a biochemist, I would be unable to add anything further to their work.

So I prefer to start with what I think are the basic principles whether or not they conform to scientific opinion. When I am convinced that the key principles are correct it becomes a matter of thinking through how nature used those principles in evolution.

Then comes the really hard part! How to explain why my solutions are different from mainstream science and how to prove I am right. Without those, the book has no value whatsoever. So I may never finish this book!

# The Key Principles.

### 1) The role of the sun is as important in humans today as it was in the creation of life.

It was the energy of the sun on the correct mix of chemicals, plus possibly a little lightening that produced the very first life forms in a muddy pond. If the sun's energy can produce life and all its complexities of cells, DNA and genes, then it can also change those things.

We know we need sunlight for dopamine and serotonin so it is impossible to think that is all we need it for.

### 2) Random gene mutations are not the prime cause of evolution.

Every scientific discipline such as physics, chemistry and biology follow strict rules or as we say, laws. They are laws because we know that given a certain input, we will always get the same output or consequences.

Random mutation does not fit with that concept. Things in nature do not happen by themselves, they happen because there is a cause. Mutations could be caused by heat, damage, chemicals, food, etc, but they cannot be random.

### 3) Muscles are the driving force in animal evolution.

Muscles give every animal its shape and every animal is perfectly designed for survival. So whatever causes muscles to develop so exactly must be the key to animal evolution. The answer firstly is conversion of pain cells into muscles, then as Darwin proposed, it is 'use'. Scientists will say "No it's random mutation", but I have covered that above.

So the principle is that because pain cells all use different protein to define location, they are all a different cell 'type'. These can be converted into muscles so making the cell type unique to a muscle. Muscle use then, via the reproduction system, causes that muscle in the offspring to become bigger, stronger and more appropriate for the task of survival. And because the driving force is use, it will shape everything in the body exactly right with the best size of wings for a bird, the best shape for a beak, horses hooves, sizes of feet and fingernails.

If random mutation were the driving force of evolution some races of people would have lungs too small, or hearts too large for their body muscles, but 'use' ensures that the entire body is correctly balanced for the job to be done.

## 4) The genes of our offspring are changed by what we do in our life.

Nature has devised a mechanism for every life form from ants, spiders and birds to man that feeds important life-saving information into our genes.

An ant with a tiny brain knows exactly what it is required to do, not from its tiny brain, but from what its genes cause it to do. This is probably the most important process of all in evolution because if the genes did not secure the survival of the insect, then man would never have evolved, irrespective of muscles and their use. The same is true for plants but the process is a little more difficult to understand.

## 5) For two different cell types to fit together and bond so that they cannot be pulled apart when a muscle causes stress there must be a single strand of DNA that is used by both cells.

This means that there must be a sharing of energy between cells that is used by the same DNA code of both cells in their growth process and a unique protein that acts as the energy carrier and glue. Without this, cell growth cannot be said to be complete as there is not 'fit' with neighbour cells.

So it is with these key principles in mind that the book goes forward to expand on them with some clues and then theories on how life evolved.

Scientists will stop reading at this point and throw the book into the file marked 'trash' because none of these principles feature in their accepted ways in which the animal body works. My view is the reverse in that because none of the principles are in their understanding, there must be something wrong in their conclusions and I suggest the cause of their misunderstanding later. But please remember that I do not claim that everything in this book is correct and I offer no proof other than logic as I do not have a microscope or a dead mouse.

# Chapter Three

## THE CLUES TO LIFE AND EVOLUTION

The purpose of this chapter is to suggest that there are clues to how animal life evolved in,
1. The different levels and quantities of energy available on Earth.
2. The effect of energy on plant growth.
3. Animal instinct.
4. The use of muscles and senses by animals.
5. Cancer.

*And conclusions can be reached from the clues that electromagnetic energy from the sun is carried by sodium ions through the nerves and into cells, the brain and the reproduction system to enable cells to grow, muscles to contract, memories to be created and stored as strands of DNA in genes, then passed to offspring to assist survival of the species.*

The processes and mechanisms that animals and plants have evolved to deliver the answers are discussed later.

## The clues from natural energy.

To understand life one first has to look at everything that nature had available in order to construct something as complex as a human. All life forms are a massive range of different cell types and so one has to ask how, were so many cell types possible?

Nature had the availability of the many different wavelengths of the sun, each a different level of energy. It had a range of chemical elements; again each of a different level of energy. It had further unique wavelengths from the moon and planets, and gravity plus sound waves, another form of a level of energy. Adding to this complex list, each of these various levels of energy was available in different amounts and at different times of the day or year.

So one can immediately conclude that inevitably all these sources and levels of levels of energy must have been used to produce such things as genes, motor neurons, signals to trigger muscle action, hearing and even the type of natural reaction to stimuli i.e. personality.

An electron is of no value to an animal, or photosynthesis in a plant, by itself as it is just a vehicle for transporting energy. Whilst a cell membrane, which is made up of a mix of unique proteins as a 'gate', will allow sodium ions to take electrons into a cell when the cell voltage is low, and potassium ions will take the electrons out to restore the cell's normal rest potential, this is simply a mechanism for placing new energy of the correct type into a cell. It is the energy that the electrons are carrying, in the form of an electromagnetic field, that is its value to an animal or plant, and this energy can only come from the sun. So I will use the term 'protein gate' in the rest of this book.

Whilst the term Action Potential may be relevant for muscles, for other purposes it merely reflects the number of electrons that are used in the transfer of a packet of electrons carrying unique energy levels into a cell for growth, and this will vary according to the number of wavelengths of energy from the sun that are required by each cell type.

So one can draw simple conclusions that, for example, each muscle uses a different energy level on its electrons, and number of electrons (Action potential being a 'packet' of electrons) to trigger its use and this level comes from the sun, to the motor neurons and each has been set to fire the packet of electrons with a specific and unique energy level down its nerve axon. This is known because each signal travels at a different speed, and speed is due to energy. The conclusion is reinforced by the hundreds of pain sensors that pass their signal up the spinal column to the brain in just two pairs of thirty-two nerves, each with a bundle of axons. Electrons carrying their energy to the brain must be carrying a unique level of energy set at its protein origin for the brain to be able to identify the source of pain. I will discuss the process further in chapter four.

As the prime source of energy is the sun, logic suggests that each neuron and motor neuron must have captured and become adjusted to specific wavelengths of the sun and have, in turn, converted these wavelength into a specific levels of energy on the electrons in a nerve that are passed between ions.

That conclusion suggests that a motor neuron is very similar to a brain neuron that captures a memory. The only difference is that a motor neuron energy level is probably inherited whilst a brain neuron need not be. The brain of a baby simply has to learn which motor neuron triggers which muscle and store it in the brain neuron. Searching for a particular memory is similar to searching for the correct motor neuron to move the required muscle.

One can further suggest that each DNA strand must be a unique level of energy and each gene a combination of these energy levels because the unique protein molecule that each gene produces is unique. Protein is a unique carbon molecule that matches the unique combination of levels of energy of strands of DNA. One can then assume that each cell itself is therefore created from a combination of various unique levels of energy. This is logical because the muscles that are triggered by a unique level of energy are the devices that drive the entire shape of an animal, including bones and tissue.

Because of the limited number of wavelengths of the sun and the vast range of cell types required for further evolution, nature would have had to make some adjustments and perhaps use one master gene to control a group of lesser genes in such a way that a key function such as a finger could be controlled in its overall shape yet provide the facility for the finger to evolve different aspects such as a nail or a claw.

We can conclude from all this that DNA and genes are simply 'adjustable blueprints' or 'memories' of how to produce a cell, and the 'type' of cell it creates depends on the level of energy supplied. If a gene is given the right energy supply long enough for its lead DNA to turn on and take control it will start to divide its cell, and the DNA of its surrounding cells will ensure a fit with the surrounding cells.

And so logic suggests that the basic process for animals is,

> The sun originally produced neurons (now inherited).
> Neurons originally produced DNA and genes. (also inherited).
> Genes produce cells.

And that raises the question – how can the sun enter the body to do all this? Logic suggests it must be through the eyes.

*To summarise the clue, nature has used all the wavelengths of the sun to produce DNA and genes, and thence the large range of cell types it needed to evolve into each type of plant or animal.*

None of this is planned or organised it is simply a set of chemical reactions that must inevitably occur when various levels of energy strike various chemical elements and in doing so create various molecules containing that specific level of energy, and these molecules become the basis of life.

But what is remarkable is that nature has gone one stage further.

*Whilst the sun was the prime source for defining the type of gene and thus the type of cell in plants most suitable for its environment and stable evolution, nature has developed its own store of energy from the sun, via the eyes, in neurons and motor neurons to enable the rapid and successful evolution of animals, and that is the remarkable difference between plants and animals.*

But plants also have such a capability and I will discuss that in a later chapter. A study of muscle use and brain instinct can help to clarify the process that animal life uses to assist evolution, and by applying this process backwards to plants one can begin to see the origin of the process that animals adopted to accelerate their evolution.

# The importance of time.

Each type of atom has a fixed number of electrons orbiting the nucleus and each of these electrons has a nominal radius of orbit. These nominal orbits were fixed when fusion occurred in stars millions of years ago, and they cannot be changed. The orbit radius however can move outward or inward a small amount depending on the amount of heat (energy) carried at any time.

Two or more atoms will only mix to form into molecules if the number and nominal orbits are compatible. By adding heat, the minor changes in electron orbits will occur and these are known as 'quantum leaps', and these minor changes allow some molecules to form that otherwise would not form.

So it follows that if atoms heat up in the sun, different molecules will form and these depend on the amount of time that the atoms received sunlight within a day, after which they will cool again over night.

But each electron in an atom has a different nominal orbit radius and that means that each electron responds only to a specific wavelength of the sun. Thus both the time that the sun is shining, the wavelength of the sun's rays and the heat that they contain, are critical to the formation of molecules between two or more atoms.

Now the wavelengths of the sun do not vary significantly from the tropics to northern territories, what varies is the density of the waves, the amount of atmosphere they have to pass through and the length of daylight each day. So logic shows that the molecules that form easily in the tropics will not form at all in Scotland, and molecules that form in the summer sun in Scotland will not form in shade or in winter.

*Thus we can conclude that the wavelengths of the sun and their density at a location, plus the time of daylight each day, all affect the type of molecules that are created within a plant. e. g. DNA.*

At this point just note that because the temperature inside an animal is held constant at that of blood, such molecular changes cannot occur. The only place where such changes are possible is at the eyes, where the sun shines into the body directly irrespective of body temperature.

Also note that once a molecule has been created it cannot be undone in normal circumstances (I.e. it requires the addition of more chemicals). So a molecule that formed through the eyes of an animal is fixed forever, (in normal circumstances), and I will discuss later that the molecules were formed millions of years ago and became inherited in every following animal form, including man.

# The clues from plant growth.

When the root of a plant grows underground horizontally it eventually grows upward and reaches ground level and into sunlight. Immediately the root becomes a stem and then a leaf. This tells us that it is the sun that causes genes to control or not control, or to turn on or not turn on. But it is not the heat of the sun it is both a specific wavelength (one that matches the gene controlling the root or the gene controlling the stem), and it is also the number of waves from the sun reaching a single point on the plant each second. The wavelength decides which is the controlling gene. The number of waves decides when there is enough energy for photosynthesis to work.

A plant that grows at the Equator will not grow in Scotland because the energy available is insufficient to enable the gene to start to control and photosynthesis does not start.

But for a plant to be able to adapt to changing climate or location from scattered seeds, there has to be a mechanism within each plant that measures the energy at each wavelength (I.e. wave density and time) and this becomes the threshold energy to start a gene working. So there must be a process for the plant to measures the output of the sun annually and adjusts its energy requirement accordingly. (This, I believe is one of the reasons for 'gene mutation' and it is caused directly by the sun, and the plant adjusts itself).

*To summarise the clue, both a wavelength and an amount of energy at that wavelength are required for any gene, including ours, to start to control cell growth. If plant seeds are blown to a location where the sun's strength is different, the plant will slowly change its DNA and adapt.*

# The clues from animal instinct – The most important clues

A bird knows that it must migrate to survive and it even knows the route it must take. A bird has never seen a nest being built but when it returns from migration and finds a mate, it knows exactly what to do, and each species has its own design of nest that has been handed down. A cuckoo has developed the mistaken belief that laying its eggs in other bird's nests is essential for survival. (I would put man's belief in God as an instinct in that same category). A spider knows how to build a complex web and all of these things are more complex than can be learned quickly from parents, and certainly not by trial and error as all nests and webs within a species are basically the same.

So there is definitely a passing of knowledge from one generation to the next and I believe that this comes via the fertilisation process. But 'knowledge' is perhaps the wrong word as it is more of an uncontrollable desire. A spider is probably not sufficiently intelligent to know why it should build a web but it simply has a desire to do so.

Sheep are given a mental map of the area in which they live showing where to shelter from a storm, where to get key items of food etc, and farmers believed that this map is provided via the mother's milk, but I cannot agree, firstly because the milk would be digested and the map lost, and secondly because birds and spiders do not feed their offspring milk and the process must surely be the same for all animals.

Ants are interesting. It seems to me that everything they do is instinctive. I do not believe one ant has either the intelligence or communication ability to teach another ant what it is required to do, let alone the thousands within their community. Their entire success has depended on memory built up over thousands of years and handed down automatically from generation to generation through the genes. An ant does not need to think and probably is unable to think much, instead it is pre-programmed with a desire to search for food, help to build a nest, probably without knowing why, and fighting off invaders. The difference between life and death has no meaning to an ant. It behaves as a robot and simply gains satisfaction from carrying out the programme it was given. It is not a question of 'live or die' for the sake of the community it is a case of carrying out the programme whatever the consequences. I do not think social factors have any bearing on an ant's behaviour, the colony survives because of its high population and the fact that actions carried out by their predecessors and improved upon over thousands of years and passed on in the genes have fortunately proved to be a winning technique.

So let us assume that instinct is true and that a memory, or a strong desire to do something, can be passed to the next generation. This must happen via sperm and eggs; the common factor for all animals and it shows that there must be a process for a memory or an instinctive action to pass to the reproduction system where it becomes permanently stored again as a mix of DNA molecules in a group of genes.

If a recurring event such as building a nest requires recurring identical thoughts or the recurring use of the same muscles then these must be able to create or prioritise a gene in sperm that is later passed via DNA into brain neurons of the next generation of birds.

Logic suggests that a catastrophe could pile so much chemical onto DNA within a few days to create a memory that lasts for several later generations. This could be so strong that a person actually believes that they must have led the earlier life and it might even lead to the feeling of 'reincarnation'.

Remember that the image of a web that a spider carries in its subconscious will have taken thousands of years and thousands of generations for it to build into a meaningful picture, but each generation will have added more and more molecules to the DNA so making it a higher and higher priority and a better and better defined DNA, resulting gene and final picture in the offspring's memory.

All this demonstrates that things done during life can be passed on to the next generation through the genes, just as personality is passed on. Random mutation of genes cannot do this and so one has to question whether random mutation is really the foundation of evolution, or more precisely – is the movement of DNA within the gene pool of a chromosome that is termed 'mutation' really random, or is it caused by other factors both internally from life's experiences and externally from energy variations?

One final point on DNA. We know that all the strands of DNA in a gene are the device that decides the 'type' each cell should be, and as above, genes contain memories of things that helped survival in earlier generations, so genes are a complete hierarchy from the control of a single cell to the control of what an entire animal must do for survival. But DNA and genes are not unique to animals, they are also in plants and so it must be able to perform the same functions for plants, which in essence, is to 'learn' from what happened today to assist survival of the offspring tomorrow. But 'learning' is the wrong word; it is adaptation, and it is not achieved by random mutation, it is achieved by passing on proven solutions in the seeds for offspring.

*To summarise the clues,*
>    *a) The brain has a connection to the reproduction organs.*
>    *b) The brain can cause changes to genes and those changes can put information into the offspring's brain.*
>    *c) A nerve's ion transfer mechanism can carry complex messages and images seen by the eyes, and not just pain or muscle trigger* signals.
>    *d) It confirms that important information that has led to survival for a parent during their life can be passed to the subconscious part of the offspring's brain, where it can then be used without any further intelligent decision making.*
>    *e) And the most important clue.* <u>*Gene mutations can be caused by the animal.*</u> *So is random mutation a valid argument?*

Human instinct would seem to take a back seat at the age of about three or four and from then on, things learned each day take over. We have encouraged this change and we call it 'education', but some people seem to be capable of re-opening their ancestor's instinct genes in later life and believe that they actually lived that earlier life and have been reincarnated.

However whilst instinct – the ability automatically to know what to do without thinking - is vital for survival of a simple or very young life form, there is a problem. It could destroy the desire to improve and hence the ability to learn. I often wonder why the very first human civilisations in Africa seem not have progressed at all whereas civilisations that left that continent have progressed dramatically. Could it be that those who travelled to new surroundings and a cooler climate were forced to change their habits rapidly and to re-learn how to survive to such an extent that 'learning' became more important than 'instinct' and this change has continued to exist up until today?

It could be that the more locations of differing climate that one lives in the more it is necessary to break the mould of habit and instinct and learn new ways. It's not so much the experience that is learned at each location but rather the very act of having to re-invent your way of life each time.

## The clues from the use of muscles and senses.

It follows from the clues of instinct and the spider's web that the repetitive pattern of use of the same muscles over many generations becomes ingrained into instinct and the genes, and is passed onto later generations so that they do things essential for survival without really knowing why. However if the use of muscles can be recorded and passed on to later generations in this way then it follows that muscle use by animals is itself a critical factor in evolution.

Those muscles that are used heavily will be more enhanced in the next generation. Those that are rarely used will become less enhanced. Thus because man stood upright and had no further use of the tail the muscles were never used and the tail disappeared. When humans started cooking meat the strong jaws of the ape were no longer required and were no longer 'used', thus human's had smaller teeth and jaws.

Clearly it is a case of making the relevant genes more dominant and able to produce more growth in the next generation and vice versa. In this way the next generation has evolved to be a little better at survival.

But pain sensors tell us something even more important. Firstly, the original life forms were transparent so that the sun's energy passed right into the body to provide energy for growth, but this means that each cell would be a different type depending on the mix of wavelengths it captured. There are thousands of touch sensors around the body but very few nerves exist in the spinal column to carry their message to the brain, thus it follows that the message must be coded or the brain would be unable to detect the location and, since the sun had made all these cells to be a different type, coding such a message was simple. It was done by their unique protein.

That means that the mix of energy levels produced by each touch sensor is different. The early life form was able to decide which location of its body it wished to move and so cause the unique cell types in those locations to become muscles. And because these muscles were once just such touch cells, the same coded energy message must apply but is sent the opposite way, and so all muscles are a different cell type and use a different trigger code. No random gene mutation was needed.

That unique mix of levels of energy can only come from the sun and if the process is inherited and passed to the next generation, then the sun's energy is connected to the reproduction system in animals. So logic suggests that must be the process that enables muscle 'use' to produce more dominant genes and greater cell growth for the next generation.

*The clues from muscle and sensor use are,*
*a) That there is a link between the sun, eyes, motor neurons, pain cells and the reproduction system that, during past generations, has led to a unique signal of energy being used for each of our muscles and produced cells of different types.*
*b) These links of unique energy enable the amount of use of each muscle during the life of the animal to set the priority of its DNA that is recorded in its genes, so that future generations benefit.*
*The body has a mechanism to do this as shown in the clues on instinct.*

# The clues from cancer.

Some scientists believe that the genes that form the basis of cancer were produced by random mutation, (as well as changes caused by viruses and carcinogens) and in fact they believe all our genes were formed from mutation of existing genes, my logic suggests that nature (i.e. chemistry) does not make such mistakes. We need to define exactly what we mean by 'mutation', i.e. do we mean random mistakes, or do we mean inevitable changes due to external factors?

Consider the logic. We know that there are about two hundred different forms of cancer and perhaps dozens of genes are involved in the cause of these. A known gene usually causes cancer of the breast and the same can be said for many cancers these days. If, for example, breast cancer was due to random mutation, dozens of genes would have to be randomly mutated to create the breast cancer gene. It would mean that random mutation would have to happen on a vast scale and would be a very serious problem. We would see people with deformed legs or bent fingers but this does not happen. So random mutation does not seem to be the correct answer. The problem must be 'directly caused' mutation where some specific external factor influences the mutation.

Random mutation also makes no real sense because errors in copying do not occur at cell level, so why should they occur at sperm or egg level when the process is so similar? If mutations did exist at cell level there would be many children who would suddenly become deformed for no apparent reason.

If we consider inherited cancer genes and assume at this early stage that these are not caused by random mutation, and viruses or carcinogens are unlikely to get into the reproduction system, then what can produce inherited cancer genes?

From my earlier clues, it suggests that energy of a high level is able to travel down the nerve to the reproduction system and overwhelm some DNA and cause a change to a gene. From that we can conclude that any energy passing down nerves to the reproduction system is able to create or change DNA. And the source of the energy can only be sunlight through the eyes. I will discuss this further in the chapter on cancer.

All this indicates that the sun has a bigger role in the DNA and gene creation and cell division than we think, and because of this, it must have a bigger role in some cancers than we think.

We know that some cancer genes are inherited and that many people have cancer genes in their cells yet never develop cancer. The cancer develops only when the energy that originally changed the gene into a cancerous type becomes available in the offspring to turn the cancer gene 'on'. So the energy that goes to the reproduction system to create or modify a gene (whether cancerous or not) can also go to an individual cell to turn that same gene 'on'.

A cancer gene is unlikely to be a dormant gene from the distant past such as a tail gene or a fish fin gene or everybody would have these genes and that is not the case. So an inherited cancer gene has to have been created in the reproduction system and is a 'directly caused' mutation (not random).

Because the same cancer causing gene is always present in any specific inherited cancer, it is clearly related to the nerve link from its brain neuron. It follows that it is the specific neuron that formed the DNA and control gene for the function (in a process that I will describe later), e.g. a breast, that has a direct role in causing a particular cancer gene to be produced and to create a new cell type.

These days almost every serious illness is being found to be due to the presence of a mutated gene, but the real solution is not to find a drug that cancels the effects of the gene but rather, to find out what external factor caused the gene to mutate, and stop it. Some are known such as smoking and alcohol, others are not.

*To summarize, the clues from cancer are,*
*a) Most genes, including inherited cancer genes, are not simply random mutations, they are 'directly caused mutations', and one cause must be by energy entering the eyes to neurons and down the nerve to the reproduction system, or an individual cell, where it changes DNA so causing a new gene.*
*b) Equally, the same external energy via the eyes is required to cause the*

*cancer gene in a cell to turn on and produce a cancer tumour.*

We can begin to see from this that one of the problems causing cancer is one of lifestyle; sunbathing, nightclubs, electric lights, working night-shifts and TV – not the sort of lifestyle that apes, cavemen or even the Victorians were accustomed to living, hence the recent increase in cases of some cancers. I will return to this later.

## The role of energy is obvious from BCC skin cancer.

Basal cell carcinoma tells us more because the changes can be seen easily. When a single skin cell has been changed by excessive sunlight and forms a cancer the cancer can be a small raw patch or it can grow inwards as a hole. If no action is taken the wound will become larger and deeper as the sun's energy penetrates deeper. If factor 50 sun block, or a light-proof plaster is used to totally block sunlight, a small hole will heal completely in about a month and a large hole about six months.

I know this because I have fair skin and living in The Bahamas for five months each year gave me skin cancer, and most of it through the car window!.

Correcting these by surgery takes five hospital visits and often results in a small scar. So I decided to experiment and try to fix the cancers with total sun block or light-proof plasters. The cancers immediately stopped getting worse, then shrank and some healed completely. It is probable that the cancer cells are simply dormant and cannot grow because good skin has grown over the top preventing light from turning the cells back 'on', but I cannot really say they are dead. Some restart to grow when the plaster is removed.

Larger patches of skin cancer, eg 1/4" diameter, are more difficult to cure in this way because more cells are involved and it is hard for good skin to spread over the top of a large area, and the cancer cells cannot pass the required energy to good skin for growth because there is no 'fit'. However, the cancer is easily brought under control by eliminating sunlight, and that proves my theory. (And they were leaking blood badly when I started).

I believe that this tells us that continual energy from the sun is needed for skin cancer to grow, and when this is blocked and growth is stopped, the DNA is not changed back to its original type, it is simply dormant.

The question arises as to why is energy necessary for a simple skin cancer cell to grow? The cell growth process is well understood and it does not require external energy, but it does require a protein. Each protein is a unique type for each cell and is created, in my view, by the unique level of energy passing through nerves to the cell. The protein controls cell division because it is the enzyme used in the growth process. Thus energy from the sun is required for cell division. In

27

the case of BCC this energy comes directly from the sun on the skin. In internal cancers there must be another route and I will explain later that the route is the nerve.

The mechanism of ion transfer into and out of a cell is, as discussed earlier, a mechanism for carrying energy into a cell in the form of a field on an electron. The rays from the sun are exactly the same form of energy because they are electromagnetic fields, thus the sun's energy can enter the skin cells to cause cancer, and transfer to electrons within the body to cause internal cancers. And skin cancer confirms that cells require energy, not electrons.

In plants, photosynthesis works whenever there is sufficient sun and it puts ATP energy into the sap to provide the energy source for growth. The same is true when animals eat as this puts sugar into the blood. But adequate food, or adequate energy to process it, merely provide the energy for growth and this is made available to all cells in the structure via sap or blood. So every plant or animal must also have a separate process to decide which particular cell should grow, and when, from the thousands that it has, and for plants, this depends on which cells are in the sun at the time.

*So the further clues are,*

*Cells require electromagnetic energy for growth, in addition to 'food' energy. This energy triggers a gene to turn 'on'.*

*This energy can pass between cells that 'fit', but not if there is no 'fit', (where 'fit' must imply a matching energy requirement, i.e. the same DNA).*

*The amount of energy available dictates the speed of growth of a tumour.*

# PART ONE.

# EVOLUTION OF THE KEY FUNCTIONS OF MAN

## Chapter Four

## THE EVOLUTION OF MAN

## HOW MOVEMENT MAY HAVE EVOLVED

The purpose of this chapter covers,
- The evolution of the nervous system.
- The evolution of muscles.
- The evolution of the reproduction system

The only things available for animal evolution were chemicals – mostly from plants - and energy. No magic. No miracles. So how did an animal life form emerge? The process of evolution is best considered in very small steps and to follow the problems that life forms would have had to face in obtaining food and achieving survival.

A plant cannot move by itself because it has no nervous system. It can produce branches wherever there is sunlight, and roots where there is nourishment. In short, the plant cells themselves will either grow or not depending on the conditions where they are and so they do not require central control. If humans followed that same decentralised process we would grow a variety of legs in different quantities so that walking would be rather difficult. Central control is essential.

Animals have such a method of central control using a brain and electrical / chemical energy, so how did this originate?

Most of the cell types required for the basic function of a human were created when early pond life was transparent. The sun would have created many different pieces of DNA from its many wavelengths, and at this time there was no helix of chromosomes to provide any control. These basic cells then suffered damage in their use and cell grouping became necessary to add strength, so causing all the different cell types for tendons, ligaments, toe nails and eye lids.

The sun's energy, passing through such transparent cells would cause the energy stored in the molecules of each protein or DNA rung to be adjusted to unique levels based on specific wavelengths of the sun. This process causes electrons in the atoms of chemicals to change orbit so changing the type of molecule produced when chemicals are mixed. Different wavelengths of the sun with different energy levels will therefore produce different molecules of DNA.

I believe the junction of all the nerve pathways became located at the surface of the head and this junction either always was, or eventually became, a number of transparent cells and ultimately an eye. At this location all the various energy levels of light waves from the sun could penetrate the transparent cells and

31

enter the head without losing their energy and whilst the sun's rays have an oscillating frequency or wavelength that was critical, their important factor was their various levels of energy.

The energy in a wave from the sun is electromagnetic and this energy can 'excite' electrons, i.e. to take a quantum leap into a higher orbit in each atom.

Thus around the location of the transparent cells, it is conceivable that a whole range of energy flows from the sun would have produced a whole range of cells in which electrons in chemicals merge at different energy levels. These became motor neurons that send signals with a specific potential and energy level to each muscle.

In my opinion the evolutionary process was; first the simple spinal cord and nerves to cells; then a simple eye, blood veins, motor neurons and a simple brain; then muscles; then a whole range of lesser senses and functions leading to a complex brain, intelligence and sight.

These changes took place over a very long period of time and in a changing environment from a muddy pond, to the sea and then back onto land. It is the period long before Darwin's theories of evolution by survival of the fittest.

## The creation of the nervous system.

The development of nerves was the first and key step in changing life forms from plants to animals. Without a nerve there could not be muscles, eyes, brains or many other functioning organs. So how could nerves have evolved?

The fundamental points are,
- Cells require energy to replicate and grow (in addition to food).
- Any cell can be converted to an energy carrying nerve.
- There are two separate sources of energy. The sun and gravity. The shape of an animal (long versus spherical) depends on which was the first source used.

There are also some known facts,
1. We know that a partially damaged spinal nerve can repair itself but a severed spinal nerve cannot. This is true even for a child that is in its fastest growth years.
2. Thus it is clear that it is necessary for electrical energy to flow through the nerve to enable it to grow or repair.
3. From this we can say that growth of a nerve requires electrical energy to flow through the cell, and a nerve will grow as fat as is necessary to do its job.

So how could this have occurred during the process of evolution? How could a cell become a nerve and then carry electrical signals throughout the body.

Clearly external electromagnetic forces must have been at work and it becomes a question of whether it was energy from the sun, like solar panels, or energy from gravity, or even the Earth's magnetic field. To achieve any flow of electricity there has to be a difference in potential, that is to say one part of a life form must be held at zero potential or 'ground' whilst electromagnetic energy is added to another part to create a voltage.

As an example; if a steel spike is stuck into a jelly and voltage is applied to the spike, nothing happens. You cannot 'push' energy anywhere. But if a second spike is stuck into the opposite side of the jelly and 'grounded' then voltage is applied, a current of energy will begin to flow between the two spikes. Energy will always try to equalise itself.

Simple pond life is a life form that grows by absorbing nourishment, and dividing its cell into two, then four, then eight. It can be seen immediately that there are only two possible ways that it can grow; either flat like a jellyfish or long like a worm. This is so because if it grew like a ball the innermost cells could not absorb food and whilst cells can pass chemicals and energy to each other, there are limitations to its size.

## First, nerves created by the energy of gravity.

If you have read my first book, or my earlier summary, you will know that I have concluded that gravity is produced by trillions of positive electromagnetic waves rising up from the protons in every atom within the planet Earth, and whilst the level of energy of each wave is considerably less than that of a wave from the sun, the density of the waves is thousands of times greater, and it is the density, together with the fact that they exist both day and night, that leads me to conclude that gravity was more likely to be the creator of the first nerve rather than the sun, and the following is how I believe it happened.

If a worm shaped amoeba (I.e. early pond life) was located in a gently flowing stream, and the rippling water from a stream caused the tail to waggle a remarkable thing becomes possible. An electric current can start to flow!

One of the laws of physics is that when an electrical conductor such as a piece of wire crosses a magnetic field, a current is induced into the conductor. It is the principle of the dynamo or electricity generator.

Now I do not want to get involved in the structure of axons or myelin because these things did not exist at the time. All there was is a collection of tissue cells, but when a worm's tail is forced to waggle randomly by the flowing stream, it cuts the magnetic lines of force leaving vertically from the ground that form gravity. (Actually the mechanism is rather more complex because the magnetic

33

polarity alternates from north to south as the gravity field passes through any point upwards at the speed of light. That is one reason why a compass cannot detect the field). The free electrons in the worm will be caused to oscillate, or increase their level of energy.

This is a similar process to divining for water. I talked to a diviner from the local water company and he used a wire coat-hanger cut into two pieces. Each piece was bent $90^0$ to form handles. When water was detected the field caused the horizontal 'pointers' to move towards each other. He gave me a demonstration and it was unbelievable! My understanding is that water flowing underground across the gravitational electromagnetic field induces a current in the water. This current produces a field of its own and it is this field that is detected by divining rods.

The process is rather like the 'quantum leap' of an electron in an atom when energy is added, but in this case the electromagnetic field of gravity builds up a potential in the water that causes a current to flow.

Returning to the case of the worm, a tiny current or voltage difference is induced in the worm's tail end cells. This current is made of tiny electrons that flow from cell to cell of the tail. The front of the worm cannot waggle if the tail is waggling, or perhaps the front is secured to the stream bank to stop it being washed away, so a potential difference, i. e. a voltage, will exist between the front and the tail of the worm.

But the process is not quite the same as a current flowing through a cable because cells are not like atoms. The only way the current can flow is for the electrons to travel as ions. Each cell has a membrane in which sodium ion gates allow positive sodium ions in, and potassium ion gates that allow positive potassium ions out, in a way that ensures that the polarity of the cell remains as it should be. Thus if a voltage potential causes electrons start to move in the tail of the worm in this way into neighbouring cells it can do it be ion transfer. In fact every cell in the body can do this if a potential difference is created to cause the move.

Energy will always try to achieve stability, or equality, so that the current will try to move from the tail up to the front or 'head'.

This process may eventually change the cells into different types of cell as the process becomes used. The spinal nerves carry many different signals and so the ion gates that control electron movement would have to cater for a large number of different energy levels, whereas the peripheral nerves would have a smaller range, and the gates into individual cells, which include protein, will act as a filter and only allow a specific level of energy on an electron to pass through.

So an electrical circuit or nerve system would begin to emerge, not as a growth from the brain in order to achieve control, but as a growth from the tail because waggles produce electricity.

Because the electrical energy flowed only one way, it produced its 'head'

34

and 'tail' ends. As the amoeba worm became fatter there would be not just one spinal nerve but dozens of nerves starting at various points along its length and finishing at the head, all due to the waggling.

The reason gravity has this ability is because of the density of the waves and the continual duration day and night, whatever the weather, that the energy is applied.

Gravity would produce long animals such as lizards, crocodiles and humans, but the sun is also necessary for animals to grow fat. The peripheral nerves that lead from the spinal cord would have been started by the waggling process but this would only make them about 1mm long as that would have been the diameter of the pond-life's body at the time. The extended length of these nerves would have arisen by the sun passing energy down the spinal cord and across the peripheral nerves to cells which needed the energy to produce protein and grow. So the nerves grew and the pond-life became fat. Then, of course, the nerves became paths to trigger muscles as well as to grow cells and tissue around muscles.

## Secondly, nerves created by the sun.

Animals with nerves created only by the sun will have no backbone because there is no path in which energy can flow to change cells into nerves. Such animals will be crabs and spiders, and their ability to evolve further is held up because they become omni-directional movers, and cannot move at speed.

But cells help each other and if direct sunlight energy falls on an outer cell it will feed the energy to inner cells in the same ion transfer process described above, and allow them to grow. But because more and more energy was required, the process improved further and some of the inner cells became permanent electrical conductors that could grow as long as necessary to feed energy to the cells that need it.

Unlike gravity produced nerves that were created by a potential difference, this would be a process where inner cells ask for energy from surrounding cells rather than a push of energy from neighbours.

Thus a dome shaped animal like a jellyfish emerges. This animal can become fat, but not long. This same process of capturing energy from the sun is necessary for gravity based nerve animals as that is the only way they can become fat, but the sun's energy would have passed down the spinal column. I will discuss this again later because when furry skin is added the outer cells no longer receive direct sunlight and an alternative process emerged.

Note that the usable energy of an electron is not its voltage, it is its electromagnetic field and this energy is a variable depending on its source, and in most cases the source is the sun. One can visualise the electron movement as being like a bellows pump that opens and shuts, and is recharged by the next electron to open again. So what is really happening is energy, carried by electrons, is moving through cells to where it needs to be. The specific energy in the field will have been put there by the sun and will travel through the nerve to a cell that will only allow that specific energy through its ion gates. Thus a cell gains the unique energy from the sun that it requires for specific protein making, division and growth. The nerve may contain many levels of energy but each cell will only accept the specific energy it needs. It is part of its definition of 'cell type'.

A nerve will become as long or as fat as is necessary to transmit the amount of energy attached to electrons required by its dependent cells. Some nerves also evolved a process of transmitting energy the opposite way – to drive muscles, and I will discuss this next. The change occurred in both long animals and flat dome animals, and it was this change that enabled both types to move.

The anemone was one of the first species to have nerves and one can see that it could have been either the sun or gravity that produced its nerves. The motion of its tentacles produced by the currents of moving water could cause the electromagnetic waves of gravity to induce currents in the anemone cells so turning them into current carriers. Even plant roots produce tiny currents when touched but clearly there has been inadequate motion to turn them in to nerves.

The nervous system, the muscles, the motor neurons and the eye are of course all related and connected together but it was the nerve that had to develop first whilst motor neurons were also developing in parallel using energy from the sun, and I will discuss that later.

Meanwhile the early brain, which would have been just a lump of tissue, was acting rather like a battery, simply storing the electrons and energy that came up via the nerves, or directly down from sunlight.

## The Evolution of Muscles and Pain Sensors.

The nerves and blood veins are both two-way systems. For a single cell amoeba to grow into a worm, food had to be provided from one cell and passed to all the others, thus a single cell type became a food carrier and grew as long as was necessary to do the job of feeding a long worm. The veins are two-way because in later evolution, protein had to be passed from dividing cells to neurons in the brain to fire energy down the nerves for growth.

Likewise, neurons did not simply send energy to all cells, they received energy as pain from all cell types in the body and so know where problems were occurring.

When evolution of the spinal nerve of an amoeba was completed the voltage in the head would exceed that produced in its waggling tail and there would be an excess of electrons stored in the 'head'. Similarly, if the sun shines on neurons all day, energy is absorbed by them and stored in the head. If the stream

36

that caused the tail to waggle now ran dry so that the tail stopped being waggled, a current must flow the opposite way, from the battery in the head to the tail to equalise the energy.

The effect of this current, again flowing through an oscillating field of gravity electromagnetic waves and now able to move very fast, is to cause the tail to waggle. Such a process would mean that the worm is more likely to move to a position in the stream that is not dry, and survive, while those that have no nerve pathway and cannot waggle will die. Hence this reverse electron flow became essential for life.

The worm would discover that by pushing the same pain energy mix back down the nerve down the nervous system at any time would cause its body to move and gain access to where the food is best. Only the chosen cell would move because only that cell would allow the energy mix through the protein cell gate.

But all messages pass to the brain through the spinal cord where there are only two pairs of 31 (or 32) nerves, each with a bundle of pathways, but not enough to provide a direct link with thousands of cells, so how do messages get to the right places?

Pain sensors provide the clue. These are proteins at nerve endings and these would evolve at this time so that the worm could become aware of dangers, but the interesting point about these is that the brain can locate the source of the pain accurately. There are only two ways I can think of that could achieve this. One is to have a system of isolation that is able to close off nerves one by one until the nerve carrying the pain is identified, but that is complicated and involves devices that do not exist. The second method is to have an 'address system' where the message is also a code of location, similar to that used by satellite and cable television companies to turn subscribers off who fail to pay their bill. In this system the electrons bearing the message of pain carry the unique fields of energy equivalent to the protein of that pain sensor, or, as discussed above, the energy that the ion gates of the cell will allow through.

Whether this code is a single electron carrying a unique energy, or whether it is a packet of electrons each with a unique energy that together match the electron orbits of a complete protein molecule, is not completely clear, but to satisfy the known 'Action Potential' discovered by scientists that trigger muscles I believe it must be a packet of electrons that forms the address code.

The 'Action Potential' discovered by scientists defines the number of electrons released by the protein of a pain sensor, and this will be common to a number of unique proteins because it is just a reflection of the number of atoms and electron orbits that make up the protein molecule. But each electron will carry a different level of energy (i. e. its field size) depending on the number of each type of atom making up the protein molecules, and it is probably this combination of energy levels attached to the stream of electrons that forms the address code.

The many cell types used to identify the source of pain would have been created by the sun shining through the transparent cells, each sending its own unique signal up the nerves to the head, so the worm could identify the source of the problem.

So the conclusion I reach is that every message, including those from motor neurons that trigger muscles, is unique in its mix of energy levels attached to the electrons within the packet. Again the neuron and motor neuron cell types would have been created by the sun where each brain neuron is a store of just one unique wavelength of energy, but when several are fired at once, a unique mix of wavelengths is sent down a nerve, just as the mix of energy making up a protein in a pain cell sends a unique mix of energy up the nerve. It was because the cell growth process routinely sends protein through the blood to fire a specific group of brain neurons to release a unique mix of energy down to a specific cell that animals used the same process to 'manually' fire a mix of energy to trigger muscle action.

Pain sensors that are feeding long term pain to the brain would eventually run out of electron message carriers if there were not some way of replenishing the stock in the cells.

So at this time the cells throughout the early pond life's body would now be attached to nerves to provide their unique energy for the growth of internal cells, and the neurons that were exposed to direct sunlight would capture particular wavelengths of the sun. So the amoeba now has energy from the sun stored in neurons and motor neurons and nerves that connect to every cell, and it now has the ability to send energy deliberately to any group of identical cells, and not just for the purpose of cell growth. The effect of doing this would be to jolt the cells and cause the amoeba to move.

So it could then soon learn which motor neuron to fire to release the specific mix of wavelengths to jolt the group of cells it wished to move, and these cells became a muscle.

Note that the size of a nerve is set by the quantity of electrons and their level of energy being released by the motor neurons. In my opinion it is this that fixes the size of axons, not the other way round.

The cells in muscles rarely grow because their neurons are motor neurons and all the energy they transmit is used to trigger muscles so it is of no use to the cell genes for the process of cell division. The use of energy has been changed and it is the specific muscle signal that matches the related muscle DNA that produces dominance in genes. As new cell types became created by the merging of existing muscle cells, new strength and features could be added to the muscle so that it behaved exactly as the animal wished.

And because the original muscle cells produced all the variations that added strength, the muscle gene controls the growth and size of the entire muscle structure including bone. To achieve this a DNA memory system became necessary, but no 'gene mutations' have been required to achieve all of the above functions.

## The evolution of the reproduction System.

The reproduction system, DNA and genes could only be created once a nervous system was completed because the helix has to be copied, and electrical energy is required to flow into the system and down each helix to produce DNA of the right type and in the right growth proportion (dominance). Once a helix is produced it is passed into every cell in the body, so every cell could be a seed, but first the helix has to be made.

I have discussed neurons and nerves and the fact that they take energy from the sun and send it to muscles to trigger them, or to cells that need it for growth. It is clear from the structure of neurons in the brain that all neurons are connected together by pathways, thus it is logical that the energy in any nerve anywhere can be sent down the spinal nerves to the reproduction system, and I will explain later how this produces gene dominance for bigger muscles, and different climates. It is the method of slowly changing DNA to produce offspring with a better chance of survival. It is directly caused adjustment of genes to improve the species.

The original amoeba in a muddy pond probably did not have chromosomes and a helix of DNA as it simply divided into two, then four. But to stop doing this, and instead produce different cell types that would allow it to grow into a worm, it required the sun to change DNA as each cell divided, so producing new types. The new cell type could then be changed into nerves. Once a central nervous system was complete, the first process may have been that the tail end cells of the worm simply broke off because the thin body would not be able to support a long length but these cells would now contain the helix and all the genes of the original worm.

If many broken end cells from many worms came together in a muddy pond there is a possibility that two would become pressed together and these two would be rather like a single cell that has divided, except in this case there would be complete strings of DNA and chromosomes, whereas when cells divide the chromosomes divide into two.

Some of these broken cells would simply divide and grow by themselves but others would mix together and take the strongest genes from each chromosome leaving the others to die, so forming a new double helix.

So the beginnings of a process of reproduction using two cells from two animals emerged and the cells involved were those at the end of the spine of the worm in exactly the same position where the reproduction organs are today, except of course animals grew tails and legs that surrounded these end cells.

39

If these break-away cells were the beginning of reproduction and fertilisation then the process of gene strengthening occurred at ambient water temperature and perhaps this became the optimum temperature. As the worm evolved and developed warm blood circulation, the male reproduction organs would have been left outside the body in order to continue working at their original optimum temperature.

## Chapter Five

### THE BEGINNING OF INTELLIGENCE

The purpose of this chapter is to suggest,
- How motor neurons were created.
- How consciousness may have developed.
- The evolution of the eye.
- The evolution of senses.
- The evolution of the brain.
- The evolution of memory

## Evolution of Motor Neurons.

A neuron and a motor neuron are basically the same thing storing just one wavelength of the sun's energy, except that a neuron is triggered subconsciously and sends its energy to cells for cell division and growth, but a motor neuron is triggered consciously by choice and sends energy to cause muscle contraction. The role of energy has been changed by early animals.

Nerves and motor neurons that send signals to trigger muscles are actually very complex. A neuron can 'fire' or not 'fire' depending on all its inputs and there may be thousands of such inputs all coordinated by the brain. The reason for this is that the protein gate at every cell will only allow a specific mix of energy through, and that mix must contain every wavelength of the sun that each rung of DNA in the cell requires. Thus many neurons must be ready to fire before a motor neuron can fire. The signals are pulses that flow through the nerve and fire at least once a second and perhaps fifty times a second. It is the field size on the electron that is unique to each muscle or sense.

This can be seen in the frog where the response to an electric current in the stomach nerve requires a jolt ten times stronger and ten thousand times longer than the response to the current in a leg.

The amoeba worm may have discovered that by increasing the frequency of each push it could move the muscle faster and therefore travel quicker to find food. But, whilst I have already discussed nerves, the question is; how were the neurons that form the top of nerves created in the first place?

One example of how the body uses energy from the sun is the body clock. At the base of the brain there is a structure called the Suprachiasmatic Nuclei. It contains 20,000 nerve cells and is our body clock. It sets up circadian rhythms that seem to be due to molecular interactions within a single cell. This clock is set by light entering through the eyes to a special light sensing mechanism that has nothing to do with sight. It seems that night-shift workers are not normally able to adjust their body clock because they travel to and from work in daylight. However if they never see daylight, man-made light will eventually change their clock.

This seems to confirm firstly that the body takes the sun's energy through the eyes for completely separate purposes than sight, and that it is not just the specific frequency or wavelength of light that is important it is also the amount of energy.

Eventually man-made light, although 100 times less energy than sunlight, will eventually produce the chemicals that tell the body what time it is. Our body clock after transatlantic crossing is of course totally controlled by sunlight and we can normally adjust within 2 or 3 days.

Both types of neuron would receive energy through the eyes and structure its molecules to just one wavelength. This is not just a question of 'voltage' because when a muscle is required to work harder the motor neurons simply send a larger packet of electrons to the muscle, therefore the 'energy' that becomes unique to each motor neuron, muscle and related gene must be the strength of the electromagnetic field.

Once the motor neurons were created, any additional energy that entered any of these cells from elsewhere, such as from the brain, streams of electrons would be emitted from the cells and sent down the nervous system.

The voltage in the brain battery is available to these cells so that now it only becomes necessary to switch the voltage in this 'battery' to specific cells (a motor neuron and all the other neurons necessary for cell entry), and electrons carrying a unique mix of field strength will be emitted and travel down to the muscles. These motor neurons would have evolved one by one over millions of years in parallel with the evolution of each muscle.

So the amoeba or fish has now evolved to a degree where it can send a variety of different signals along its nervous system and by a process of learning, can decide which signal to turn on according the result it wants to achieve.

It is now just a short step to make a record of this progress in a gene as I have discussed earlier.

## A Summary of Nerves and Neurons.

The neurons are the largest contributor to animal evolution so I will diverge a little at this point and just explain why. (The next chapter covers their role in more detail). Coupled with the eye, they use cell types created by the sun to produce an animal into the perfect symmetric shape that it needs to be to survive. For a bird, they provide the best size of wings for it to fly, and a beak that does what the bird requires to survive.

First an explanation of their multiple roles.

1.  They take energy from sunlight, store it and then attach it to electrons and pass it down nerves to cells when they need it. All cells need energy to divide and grow.
2.  They pass their DNA and genes into the reproduction system.
3.  Some neurons have evolved to trigger energy to cells that became muscles, so allowing movement. They are motor neurons.
4.  They monitor muscle and sense use and pass the information to the reproduction system to make genes in sperm and egg more dominant and so produce better muscles and senses for the offspring.
5.  They monitor cell damage and start the immune repair process.
6.  They work together to use this information to improve cell structure and material so that repair becomes unnecessary.
7.  Other neurons formed the brain where their stored energy is passed between them and became thought.

In animals, the neurons detect specific wavelengths of the sun and pass that unique level of energy down nerves to cells that require that specific energy. The process is the same as for plants, but in plants the sun can reach cells directly. Nerves are not required.

The sun's energy comes in positively charge particles with electromagnetic field and is sent in small packets called photons. These can only enter the body through the eyes as the skull and skin would immediately extract the energy and simply cause the skin to get hot. But the neurons have access to this energy because the eyes are transparent.

The main purpose is to keep the ribosome charged with the energy they need to copy DNA and produce protein for new cells and growth. Ribosome can do this process in isolation, but they must eventually run out of energy if not recharged, and cell division would stop.

## Consciousness.

In my opinion the prime objective of the early brain was not to 'think', it was to monitor and later, to achieve contentment, which can be further summarised as the removal of fear. This is nothing more complex than having the correct voltage or chemical molecule at a particular group of cells.

Now in the suggestion I made earlier of the amoeba worm in a stream that dried up, the cessation in current flow in the stream and the flow of energy from head to tail would have led to a decreased voltage in the head and this may have become identified as a lack of contentment, then a risk, or fear. It would realise that something was not right. Equally it may regularly receive pain from some of its sensors, and so the brain would have developed consciousness and for the first time became aware of its existence.

It is probable that this change in voltage and contentment caused by the dry stream would cause the early brain in the worm to try to find a solution, in this case to send an electrical pulse down the nervous system to a developing muscle cell to cause movement. The worm has now learned that it can actually change its circumstance to achieve contentment and so it has developed the first stage of consciousness. It would realise that moving muscles could reduce the fear or pain, and so it began to think.

If the above is correct, and fear is one of the drivers toward thinking (as is pain and hunger), then perhaps certain changes adopted by some animals such as lizards and moths, might actually be driven by fear and lack of contentment. By this I mean camouflage. Does a moth change colour to match its surroundings due to random gene mutation over generations? Or does fear cause the brain to find solutions – perhaps from an earlier life-form, that allows colour to change until what it sees matches its surroundings? I believe the latter.

Simple insects probably have no such discontentment. As I discussed earlier, ants react instinctively, even to their own death, to protect the colony, because that is what has saved the colony over many past generations, and enabled the species to continue.

## Evolution of the eye.

The eye was perhaps the most significant development in evolution and, from all the previous sections, it can be seen that it was not for the reason of sight, it was to capture energy and so enable animals to become large, fat and have furry skin.

As discussed earlier, the body is made up of chemicals that are a mixture of atoms. Atoms differ essentially only in their number of electrons and their wavelengths, so it is logical that the wavelength of the sun will have an influence on the animal body, and such wavelengths are readily available and will influence an animal whether it wants it or not.

The muscles of an animal each use a different wavelength and that is because of the various energy levels, in the form of an electromagnetic field, attached to the electrons in the nerve. It was the attempted use of tissue by early life forms that created muscles from basic cell types and it is the excessive use of such a new muscle that creates, and makes the related gene more dominant.

The process of achieving this is done in the animal's reproduction system and of course has become inherited through the genes. It follows that each DNA, and an entire gene, is comprised of specific wavelengths of energy derived from the sun, and from that one can conclude that if an animal travelled to a more northern latitude where the sun is weaker, the DNA and genes will change.

If such movement occurs the seed pods or sperm is adjusted to match that

44

of the weaker sun's wavelengths either directly by the sun shining on seed pods, or by the sun shining through eyes and adjusting neurons. That is the importance of the eye! So all seeds in all life-forms are adjustable according to the climate so that the offspring are better equipped. They are not simply copies of old cells.

In early life forms all cells were transparent and the body was thin and small, but this structure was not suitable for larger life forms such as man, where inner cells would be isolated from the energy of the sun, and so the nerve system and eyes produced the answer.

And as evolution continued, and animals became more complex with many more muscles, new motor neurons would have to be created and used to drive these muscles. Thus even after the eye was fully developed in early animals, it was still necessary for the eye to capture more of the sun's wavelengths, and this must still be happening in man today even if it is only to recharge the inherited motor neurons and adjust neurons for climate changes, as such a process can never change once it has been adopted.

One can also see that too much light in the eyes could put too much energy into the body and actually destroy things and so perhaps contribute to cancer in the same way that sun causes BCC skin cancer. Certainly it must affect the genes in sperm for the reasons discussed above so leading to inherited cancer genes. Once energy is in the body it cannot be destroyed and must be slowly dissipated safely away. This is easily done where the energy goes into driving muscles, but not so where no muscles is involved and it is noticeable that cancer occurs in tissue that is not directly related to muscles. Cancer does not occur in muscles. One can also see that there is a mechanism in this process that will lead to the character factors that are followed in Astrology.

We also know many similar things; light is necessary in the eyes or we get 'SAD', and light is necessary to produce Serotonin to turn the brain on in the morning and Dopamine to trigger motor neurons, so it follows that the sun's energy is needed for the brain to work.


So returning to the historic stages of evolution, I believe transparent cells in the head of early life forms that were held above the water line in a muddy pool, were used to capture all the wavelengths of the sun and, in the same way that I suggested in 'clues' that an image involving hundreds of neurons is stored in DNA and passed to offspring, so a single neuron can become inherited to operate at a specific wavelength. And when most of the neurons were complete the transparent cells became less important for later species other than to re-enforce the energy in its neurons. But because the worm was now able to move itself with muscles it developed the transparent cells to have a retina of cells that could interpret any wavelength to form an image which could be used to identify the location of food, and so sight and movement had a joint purpose.

My guess is that the transparent cells were combined eyes of many segments that eventually split into two eyes, one on each side of the head.

The brain of some animals became able to focus the image in just two segments in such a way that three-dimensional images or 360 degree images were produced. The benefit of this became so great that two separate eyes became the preferred design that led to survival and hence we all have two eyes.

Ears probably followed a different route and were much later. It is well known that human ears developed from the movement of tiny bones at the base of each side of the head. Sound is vibration of these bones and eventually the lobes of the ear developed to capture more vibrations. So we have two ears because it was beneficial for the identical bones on both sides of the head to be able to capture these vibrations.

## Evolution of the Senses.

The senses are related to muscles because without them there is no reason to move. So the sense nerve channel probably transmits energy signals into the same part of the brain that sends signals to activate the relevant muscles.

According to scientists there is a store of energy, held in the form of protein, at the end of pain-sensing neurons. When a painful stimulus is felt, the protein excites the neuron causing it to send an electrical signal to the brain. So the particular protein involved causes a particular level of energy, or unique signal, to pass via electrons to the brain. (One can assume that all neurons work in this way and so each neuron responds to a specific protein or specific level of energy).

Applying the same logic as above, any energy of any kind added to nerves in the skin, or elsewhere, would cause their protein to release electrons up the nervous system with an energy code in their electromagnetic field equivalent to the various energies in the electrons of the protein. This added energy might be a push, a touch, heat, or a sound etc, so that they become senses. The brain can receive the coded message, de-code it, feel the 'pain' or 'sensation' and know from where it originated.

## Evolution of Brains.

Man has 100 billion neurons in the brain. These are simply devices to receive, store and send electrical energy, and I have said that this energy comes from the sun through the eyes. The original purpose was to pass energy to internal cells so that they could grow and the animal could become fat without all its cell having to receive direct sunlight. They were then used to sense pain and trigger cells to become muscles. From this it can be seen that it is a short step from deciding which muscle to trigger to move in a desired direction, to passing energy between neurons in order to decide which direction to move.

But there is another stage that evolution passed through after the helix and reproduction system were created in order to learn how to think. Some of the energy in the brain is inherited from the parents in the form of instinct and placed directly into brain neurons. This energy takes the form of electrical signals stored by the parent in groups of DNA that form a gene. When the offspring sees an image that matches an image stored in its neurons and put there from DNA in the parents, the offspring automatically know what to do, and initially probably does not really know why. But it soon realises that actions produce survival and it is then just a short step to learn to decide for oneself what to do. I have discussed this key evolutionary process separately in chapter nine.

Once thinking has started, the more the neurons are used to think and store images and thoughts, the more neurons will be created in each later generation from ordinary cells to do the job. So man's head got larger and larger through successive generations as he thought more and more.

We can begin to see why education is important. The more we use the brain, the more cells become available for our offspring, and so children of well-educated parents will be better equipped to learn. Equally, the more we use the brain by thinking, the more pathways become connected linking the neurons together. We all have similar but not identical genes therefore we all have the same compartments in the brain and the same ability to be intelligent, but the number of neurons and the size of each compartment depends on use and this will increase from generation to generation through the genes.

## Evolution of Memory.

I discussed in the section on 'clues' that an ant is like a robot and knows what to do because the instructions have been learned through centuries and handed down in genes, so these genes and their instructions are a brain with memory.

Such instinct shows us that memories of images and muscle use can be passed down nerves, stored as molecules in DNA and later passed to offspring as electrical signals directly into brain neurons. So there is a circular process of memory in life; what parents see as important for survival passes through their genes and into the memory of the brain of their offspring to assist their survival. Clearly memory stored in a brain must be just an alternative mechanism to memory stored in DNA. One takes the form of molecules, the other electrical signals. But the brain must have begun to develop before instinct existed or there would be no messages to pass on to offspring!

If we return to the worm shaped pond life that I described earlier I suggested that the rippling water in a stream caused electrons to flow towards the motionless head of the amoeba, so creating a nervous system. The electrons reaching the head formed a store of voltage that is now at a higher potential than lower down in the nervous system, so when the water stopped rippling, as it would in a drought, the electrons must flow back down the nervous system towards the tail, so causing the worm to waggle and move through the water. This mechanism enabled it to survive a drought when other amoeba could not. The tip of the nervous system became the start of the brain.

In the early stages the flow of electrons back towards the tail would have been an uncontrollable action and probably an action that the amoeba did not even know was occurring. But eventually an occasion would arise when the circuit became disconnected and then re-connected and it is this action that would create 'awareness' and lead to the amoeba's realisation that it has the ability to control movement. It would learn that closing the circuit caused the body to wiggle and enable it to swim, and leaving the circuit open would cause the wiggle to stop so that movement stopped.

And that simple process is, in my opinion, the basis of how the brain evolved but it is a little more complex because animals have many muscles, senses and things to remember, and then instinct evolved to guide the offspring to focus on survival.

But the number of things to remember in order to survive would have increased to a level where some sort of memory structure is required and so the second important function that occurs in the brain is the filing system. It is this system of compartmentalising in smaller and smaller files, right down to the smallest element that we wish to remember, that enables the brain to remember everything.

Consider that every one of the millions or so cells or neurons in the brain are exactly the same. Each one is simply a chemical that is able to hold free electrons when such electrons are made available. It does not matter at this stage what the chemical molecules actually are; simply their function.

Now imagine the compartments. Suppose we wished to remember the number of the house in which we live. First there must be a file containing all the numbers that exist and each number is a cell with free electrons in it. Now we create a personal file. This leads to home file and this leads to details about the home until we find an empty cell and label it 'house number'. We transfer an electron into this file. Then we cross refer to the numbers file and create a permanent link or pathway directly to the number of the house.

To remember the number we 'think', but it's a fast automatic process. We go into 'personal', then 'home', then 'details', then 'house number'. We open it and find an electron pops out that has come across the pathway from the numbers file. We have found the number.

The same searching process occurs when we cannot put a name to a face. We search the old school file, the local people file, the golf club file and eventually we find the face in the holiday file going back ten years in Spain.

So 'thinking' is a process of searching files one by one to provide options that might give the solution to what we want. But not all options are internal files. For example, when doing a jigsaw puzzle the options are external. We look for colours and shapes that match a gap in the puzzle until one fits.

To be able to remember a face or the picture in a jigsaw puzzle is of course more complex than remembering how to move a muscle or the number of a house. It requires sight and sight is a group of cells forming pixels in sufficient density to create a likeness of a place or face. Each pixel is a cell or neuron with an electron, but now shades or colours are required rather than simply black and white, or on and off.

So each electron carries the energy in its field that was the energy retained from the light that was on the object when it was seen and stored in the memory. So each pixel or cell not only has an electron added to create the memory, it also has a level of energy. Now when the memory is opened, all the pixels can be seen together and each pixel has a different brightness or colour so producing a recognisable object.

So the brain seems to be nothing more than a million identical rechargeable batteries, in which each battery is just one small compartment within a gigantic family tree of structures. Each route into a specific battery, or memory, is a complex pathway that leads down into smaller and smaller detailed areas of knowledge until the memory is found. And it is just an electron in a cell. Each 'battery' is just a nerve ending that has become a store of electrons.

Memory would have evolved from the senses. We feel pain and we see that the cause was a thorn and so we relate the picture of a thorn to pain and remember the picture. We tend to remember things in pictures.

Sight is the ability of cells to detect light and turn it into levels of energy that a neuron can learn to understand. But the brain's interpretation of sight is so fast it cannot be achieved via protein re-arrangement it has to be via electrons as in television.

A single picture memory probably contains less than 1% of the original information. It is probably transmitted through the body in the form of a bit stream of energy attached to electrons and becomes a more permanent memory in the form of a unique sequential mix of proteins.

If a gene has 50 carbon molecules making up its protein, each of these has orbiting electrons at a radius equivalent to $98.4^0$ F. A combination of neurons could transmit them as a stream of energy on electrons, so we can adjust the energy on free electrons in the form of a 1% dot matrix as they circle around the 50 molecules of a gene or neuron.

This 'bit stream' of sight data could be memorised directly although it is fast, but memory is probably achieved by re-arranging the protein mix in neurons to form a permanent memory. Such a protein mix can be converted back into a bit stream at any time in order to 'imagine' the original picture. It is probably the same process as a dream.

Like muscles and the nervous system, brain neurons would grow sufficiently in number to 'do the job' and so the more we need to remember, the more capacity to remember is produced. The more we use the brain, the more signals we pass to the sperm, as I described earlier, and so the priority of the related genes is increased and humans become more and more intelligent.

## An overall summary of how man came to exist.

The preceding chapters covered what I believe was the evolutionary process that led to man. I will just provide a brief summary of the entire process.

A simple single cell life form in a pond grew into a worm shape using different cell types created by the sun, and the moving water in the field of gravity caused a spinal nerve to be produced.

At the same time sun's energy was being collected through transparent cell at the head end and this enabled some cells to become neurons and store energy, and other cells to mutate to become nerves connected to the spinal nerve to feed the energy to inner cells, so that the worm could grow fat.

It then learned that by sending the right code of stored energy the opposite way in the nerves down to the source of pain, some cells in key areas could be made to move. It developed motor neurons and muscles. The more it did this the stronger the DNA in the genes became. The electromagnetic field from the flow of electricity in the nerve to contract the muscle caused damage to nearby cells and subsequent repair and cell interdependence enabled mutation of genes to produce more cell types such as tendons and bone and eventually strong usable limbs.

By realising that sending energy down nerves brought results, it sent energy between neurons and a thought process emerged. The neurons were capable of storing energy and so memory evolved.

At the same time the transparent cells that captured the energy of the sun could become interpreted as an image and sight was produced. Again neurons were used to interpret the image and store it as a memory.

Hearts and lungs were simply muscle devices that grew from original cell types according to the need. They enabled the life form to pass oxygen and food via blood to muscles to assist movement.

The animal could now leave the pond and crawl, seeing where it was going. The backbone became stronger by use and to meet the need of use.

It used its thought process more and more and the brain grew in its number of neurons to meet the need. The brain and skull became larger.

It now had all the motor neurons and brain neurons it needed, and all the limbs necessary to escape predators and catch food. It could come down from living in trees and walk upright. It was the first man.

The man moved north and his genes adjusted to the weaker sun, so life could continue.

He then did some dumb things and seems likely to make things worse in the future! He invented cars and aeroplanes. He became obese. He travelled back to the hot African sun even though his genes had adjusted to the weaker northern sun. His skin got cancer and the bright sun in his eyes overwhelmed the cells in his body that could not utilise the energy to work muscles. These cells became cancerous. He discovered tobacco and caused his lungs to deteriorate causing the cells to become cancerous. He then made so many machines that consumed the planets natural energy that he polluted the planet. The $CO_2$ produced blocked the sun's rays so that heat could not disperse out into space and the planet overheated. Every life form on the planet died, including man. Some tried to escape to the moon and other planets but the distances were too great and water was scarce. The species disappeared.

So that is all I want to cover on how man evolved and I did not mention the word God once! Every animal can develop the bodily functions it requires and can discard nerves and tissue that does not exactly fit with its requirement. Each function, (a finger, a hand, a wrist, etc ) is controlled by a single gene, and so the function can be passed onto future offspring in its entirety via the reproduction system. The more each muscle, sense and neuron was used the more dominant became the genes that were passed on to the next generation and the more perfect the design of the animal became to meet its chosen lifestyle, and equally, if a muscle no longer has a purpose and is not used, such as the tail, it disappears,. So 'man' really designed himself.

# PART TWO

# THE PROCESS OF GROWTH AND ADAPTATION.

## Chapter Six

**THE ROLE OF THE SUN'S ENERGY,**
**THE PROCESS OF CELL DIVISION AND GROWTH,**
**CELL TYPE CREATION AND MUTATIONS.**

## First, some school level physics and chemistry.

It is important to understand wavelength energy. Each wavelength, from ultraviolet down to infra-red is a specific amount of energy. Call that its 'level of energy'. But over a period of time as more and more wavelengths from the sun reach an object, more energy is delivered. Call that the 'quantity of energy'. This is of course why an object will slowly increase in temperature when left in the sun.

Now as discussed in 'clues', each gene responds to a mix of levels of energy because of its unique mix of chemical molecules, but each also contains a quantity of energy at that level because of the number of molecules of its unique type in its structure. For a cell to grow, and so copy such a gene blueprint, it requires both the correct 'level' and a sufficient 'quantity' of energy from the sun. That means it requires the correct wavelength from the sun over a period of time.

To explain the chemical process it is easiest to recap on the structure of atoms that make up chemicals, then briefly the animal process and then go back to the plant process.

Electrons orbit the nucleus of every atom. Each type of chemical has a specific number of electrons in orbit. Whether two different chemicals can mix together to form a molecule depends on the number of electrons that each chemical has in its atoms and also on the specific orbit radius of each of their electrons.

The nominal orbit radius of an electron was set by the level of energy (wavelength) that it was carrying when it was captured into the atom millions of years ago during star formation. Applying heat energy causes the electrons to move orbit to a very slightly larger radius and this enables two different chemicals to combine into a molecule that could not happen at a lower temperature.

When two chemicals are able to mix, the orbits of the electrons partially merge leaving some redundant and these are called free electrons. Also the ratio of atoms of each chemical can adjust themselves so that the number of electrons can merge into a molecule, I.e. it does not have to be one atom from each chemical, it can be six atoms of one chemical and two of the other. Or it can involve three different chemical and their atoms in various combinations.

The point is that if the right type of chemicals exist and the temperature is right, the chemicals will merge into molecules. If the temperature is wrong, they will not merge. And of course as we are discussing 'life', the chemicals must be the ones that are necessary for life.

A final point is that all electrical energy is carried by electrons no matter what the chemical composition of the material involved, and all the energy carried is contained within the electromagnetic field.

*An electron is like a bucket. It has no value to the body by itself, but it is a carrier of energy and what matters is how much energy it is carrying in its electromagnetic field. This energy is invisible and cannot be seen under a microscope.*

In life-forms the electrical energy in nerves forms into protein and vice-versa and so each protein is a mix of specific levels of energy that correspond to the electrical levels of energy passing through the nerve.

The reason for going through that again was primarily to show how proteins are made and how there can be so many different ones, and how DNA can have a different mix of atoms.

The sun has exactly the same construction of electrons orbiting the nucleus of every atom and the nominal radii of all the electrons produce a complete spectrum of colours of light radiated to Earth, but there is a difference. The sun is going through massive fusion causing very high temperatures so that all electrons in orbit are at their widest radii and the light waves that they emit are 'hot'. I.e. the electron wavelengths are shorter and contain more energy than that of their nominal orbit. Thus when all the many 'hot' wavelengths of light in the sun's spectrum strike the atoms in life on Earth their electrons jump to wider orbits. They become heated and so enabling atoms to combine into molecules. All that is required is time for all electrons to jump to wider orbits so that all atoms can merge into molecules. Once combined they cannot be undone.

Note that this effect is not just one wave with one wavelength from the sun, and one electron in an atom in a life, it is thousands of waves, all at different, but hot, wavelengths from the sun, and dozens of electrons, also all at various nominal wavelengths or orbits in atoms in the life form.

If we now discuss animals, in all cases the blood is controlled to a specific temperature, and we know that if there is any variation in this temperature of plus or minus a few degrees, the body chemicals are unable to work properly. We become ill.

*So we can say that both the wavelength of energy (i.e. the electron's orbit radius, or the field carried by free electrons in the outer shell) and the time it strikes an atom (the temperature attained) are necessary for chemicals to merge into molecules, and that means both of these factors are necessary for life processes to continue. Once the molecules are established, the constant temperature within an animal ensures that their role will always produce the required result.*

Considering these points in plants, we can see that the sun is the source of the wavelength energy and the location of the plant and the local climate (north, south, light, shade) are the factors affecting the time and the density each wavelength of energy is received I.e. the quantity of energy or heat. Once the molecules are established, the temperature of the water or air must be within a certain range or these correct molecules cannot do their job properly inside the plant, and the plant will die.

## The cell growth process and energy.

Cell growth in animals is all about energy being passed from the collectors – the eyes and neurons – and passing it to cells that are becoming weak and losing their grip with their neighbours. The energy is an exact wavelength of the sun and, in the body, takes the form of an electromagnetic field attached to an electron. It is a unique level for each rung of DNA. And because a gene is a collection of DNA, a packet of electrons forming a unique code can be passed down the nervous system

In child growth (embryonic) the process causes neurons to send a specific sequence and unique energy level down the nerves. The code is simply the wavelength energy of the DNA in each gene, and the sequence is probably simply that of the order of the genes on a helix in the brain. This triggers specific genes to turn on because the energy code exactly matches that of the DNA in the gene. The total of all the DNA ensures that a specific cell type grows, and all cells grow in the right sequence and in the right locations.

After puberty, the same coded packets of energy will be sent down the nerves, but because body building is now completed, the energy is used to produce spermatozoa and to replace cells that are beginning to lose their energy.

In adults, the energy stored in the DNA, which is like a battery, will become exhausted by the task of holding the cell's molecular bonding with surrounding cells to retain the strength of the body, and eventually the grip will weaken. This bonding is achieved by the existence of a balanced voltage and energy between neighbouring cells which, if not maintained, would cause the animal to fall apart under stress. When the cell's energy is exhausted the molecules must re-form into different types and the electrons are used up in the new molecules, so reducing the cell's potential. A new cell is required to replace the exhausted old one, which dies. The low voltage in the dying cell allows sodium ions to move into the cell. The ions carry energy and the protein gate ensures that only energy of the right unique code that the DNA requires is allowed in. Thus a cell is able to copy itself and restore its DNA.

The ribosome starts the cell replacement process by using the coded energy available from the nerve that matches the DNA in the gene that controls the cell. It may be that protein is released into the blood to cause the brain neurons to release the correctly coded energy, but I believe this is only required for cancerous cells where the correct code sequence does not exist on a healthy helix. When the new cell is complete, the energy balance is restored and growth stops. The nerve process is exactly the same as 'pain', but is in reverse, and the growth is exactly the same as embryonic growth when the cell was first produced.

So one could conclude that the ancient Chinese were correct. There is a flow, or an essential existence of energy passing between all cells in the body that hold the cells together, and without it the body has a problem. And this flow is supported by energy from the nerves, but is in addition to the flow through the nerves.

The pain system is probably an evolution or adaptation of the growth system, the ultimate purpose in being able to jolt the muscle and get away from the cause of the pain. But this commonality means that the protein in the cell that sends a pain signal to jolt the muscle is the same as the signal sent by the brain when the animal wishes to move. That means every cell within a single muscle structure contains the same muscle protein in addition to its own unique 'cell type' protein.

Thus growth, or cell division, is controlled by the muscle gene and a muscle gene controls growth of its total structure of cells, tendons, ligaments etc. This is not surprising as every cell within the muscle structure evolved from a single cell by using the muscle and its signal, and it ensures that the entire muscle structure grows in a balanced way whilst allowing each individual cell to reinforce its contact with its neighbour in the event of strain of ageing.

In muscle cells the neuron growth energy is used consciously by the person with motor neurons to trigger muscles contraction and therefore cannot be used for growth of the actual contracting mechanism of the muscle. Muscle cell increase or growth can only occur through adjustment of seeds / sperm for the next generation.

Muscle use therefore places a lot of energy from the motor neuron into the muscle cell and this must be reversed to restore an energy balance. I believe that process occurs while we sleep because energy can only change direction when the brain is turned off, and there is possible evidence for this in 'rapid eye movement', as masses of energy return from the muscle to the motor neuron, where it is stored for use next day. No energy has been lost in the process. Perhaps that is the purpose of sleep?

So in normal growth all the energy returns to its starting place and the body remains in balance, but cancer is different. Because a single cell can never become complete, the neuron is caused to 'work overtime' and send masses of signal energy to the cell, causing more and more protein to enter the blood and more cell division. But the body has no way of restoring such a large amount of unique energy to a single neuron. Dopamine triggers motor neurons but the motor neuron energy level is inherited to a specific level it cannot be restored by

Dopamine, however muscles do not have a problem because the energy is fed back during sleep. The only way a neuron can be restored with energy for cancer growth is by the sun shining through the eyes, and I will discuss this in the section on cancer.

It was the sun that originally created each neuron with its unique energy millions of years ago and although the retina is now used for sight, the connections to neurons must still exist. In fact I believe the eyes still have a role in adjusting the DNA to match the wavelengths of the sun in a new location, and I will discuss this later.

Plants, of course, do not divide and replicate cells because they use embryonic growth to grow new leaves every season rather than repair damaged old ones. They have no nerves and no immune system and their growth process is based on molecular energy not electrical energy.

## Cell division.

The main purpose of a cell is to stick to its neighbour to form a structure and this is normally a structure that supports a muscle. This adhesion is probably a molecular function and it this molecular flow that becomes weak over time as energy is inevitably dispersed.

That means that either the DNA in the cell is losing strength or the total energy in the cell is losing energy. My theory is that the 'lead' DNA is like a molecular battery that releases energy to produce a bond with identical sister DNA in a neighbour cell so holding the two together, but when the battery gets low a whole new cell with new DNA is produced, charged up with energy from the nerve (and sun) to produce a new bond.

I believe this weakening of the bond and reducing energy level is detected within the cell and the ribosome is caused to start the cell replacement process. But of course there are genes higher up in the total helix structure that may need to be in a state that allows cell replacement, and that is a complex subject, so I will stay with the single cell.

DNA is the blueprint of how a cell should be made. Each DNA is a specific level of energy and all the DNA in a cell produces the specific type of cell. The cell 'asks for' a specific protein that matches all of its DNA and it is the ribosome role to provide it.

The cell division and replacement process now starts and I have provided the complete process below.

The unique code of energy used to start cell division and DNA replication, is the same as the address code that I discussed in chapter four on pain sensors. The ion gates and protein in the cell membrane ensure that only an electron that is carrying the correct energy code can enter a cell when the neurons will actually be full of electrons with dozens of different energy levels.

Plant cells are dividing and producing growth all the time and energy of exactly the right wavelength is required to do this, for each gene and cell. There is no point in starting photosynthesis if the sun is weak, and so there is a preliminary step in growth in which energy (directly or indirectly) from the sun of exactly the correct wavelength for the gene controlling a cell, must strike the cell to produce the unique protein required as an enzyme for growth, so causing the cell to start to divide. This step then ensures that photosynthesis will be successful. If the step did not occur, photosynthesis might fail, and the cell would die.

The following pages provide a simple overall summary of how energy, nerves, protein and DNA all work together to produce cell division, DNA replication and growth. And one can begin to see from this some aspects of the process that may be helpful in preventing growth of tumours.

## A simple summary of the growth process in adults
(Excluding those parts that are well known already).

- The only action required in the body of an adult is maintenance of the adhesion between every cell and that means cells must occasionally be replaced through cell division and replacement. (If this were not done then things like fingers would fall off when we held something tightly!)
- Remember that DNA, proteins and cell type are all inherited and so it is just a matter of energy to start the process, and it is the same energy that cells commandeered to become muscles.
- Muscles cells do not grow because the energy in the nerve, produced by special motor neurons produced millions of years ago during evolution, fire a specific mix of energy, (the fields on electrons) that is used to trigger the muscle into action, but the triggering passes this energy to the reproduction organs so that the next generation has the correct gene priority leading to optimum size of muscle cells in the offspring. The animal adjusts to optimise its strength based on 'use' from parent to child. This is covered in a later chapter.
- All cells in an animal are controlled by the DNA of the related muscle except for the cells involved in producing offspring such as the breast, prostate and reproduction organs where there are no muscles.
- Every cell relating to a muscle contains the muscle DNA so that all surrounding cells grow in the same proportion as the muscle.
- But all the surrounding cells are of different types and, for their unique DNA to be turned on during life so that the cells can divide and grow, they must receive their own unique energy to turn on their unique DNA.

- The ribosome exist in each cell to produce the unique protein using all the DNA that forms their cell type as a guide. Some of this is passed into the blood stream and flows to the brain neurons where the protein energy is matched to neurons, and they fire energy down the nerve to the cell.
- The energy is a unique code of several wavelengths attached to electrons that is an exact match with the unique protein and the cell's DNA. There will probably be the same number of electrons as the number of DNA strands in the cell. Each cell has a protein 'gate' that only allows the correct code of energy through from the nerve.
- So the cells around the muscle receive their unique code from the nerve that they need to turn on all their DNA and start the DNA copying process, leading to cell division and so support the amount of growth required by the muscle DNA that was set by the sperm and egg when the offspring became fertilised and grew.
- So now every cell around a muscles is able to grow at the same rate as the muscle (dictated by the parents) even though they are all different types.
- It can be seen that protein is needed because animals have a nervous system where energy is taken in through the eyes, instead of as direct sunlight onto cells, as is the case in plants, (and simple BCC skin cancer) and because so many different cell types are involved, the energy supplied is a complex mix of wavelengths and not just one wavelength.
- One can see this process working in a simple virus. When a virus enters a body it becomes blocked from all daylight and cannot survive. Its solution is to steal human DNA and the energy that goes with it for itself. The virus can now survive. It has mutated and will continue to grow from human nerve energy.

..............

- There is a second growth process used in the case of injury. Where a wound must be filled, the immune system releases stem cells. The stem cell's 'type of cell' is set by molecules crossing from nearby cells causing the stem cells to select the energy that correctly matches the DNA from their pool of all DNA. That will include the original muscle DNA that all nearby cells must contain. The ribosome now goes to work and produce the unique protein which then allows the brain to fire the correct code of energy so that the new cells divide and grow to fill the wound.
- One can see from this process that changed DNA that has become a cause of cancer can be prevented from growing as a tumour by a) Filtering the blood somehow to remove the bad protein that produces the energy the cancer DNA requires. b) Wearing special glasses to prevent the unique

61

wavelength required by the cancer from ever getting into the body. c) A drug injected into the blood that breaks down the offending protein into its basic parts. d) Inserting a steel needle into the nerve going to the cancerous cell to disperse the unique energy.

- It can be seen that if a carcinogen causes a cancerous change to DNA it will be a while before the unique protein is prepared, sent to the brain and energy of the new cancerous type is sent down the nerve to start cell division and growth. So lung cancer can occur sometime after giving up smoking.

- It can also be seen that exercise will, to a small extent, help to slow the growth of a tumour because the energy needed for the muscle DNA to be turned on in the cancerous cell is burned up so preventing, or at least slowing, the cancerous DNA and cell from doing anything.

- The protein gate in each cell is possibly there to stop the wrong energy entering the cell and changing the DNA (remember sunlight created DNA millions of years ago). Thus it stops cancerous growth. If excessive high UV energy gets into the nerves it may overwhelm the protein gate and change the DNA and that, I believe, is one cause of breast and other non-muscle organ cancers, where there is no muscle in total control and no way of dispersing the energy.

## The role of the sun in gene creation and improvement.

From the 'clues' discussed in an earlier chapter, each strand of DNA matches the energy of one wavelength of the sun and each gene and each cell is a mixture of these specific levels of energy. And we can conclude from the clue on cancer, and from the discussion above, that for a gene to be able to produce the exact type of protein molecule that is required for the cell type to grow, energy of the correct level carried by electrons, must pass to the gene for a sufficient time.

We can deduce more conclusions from this chemistry:-

A gene or DNA will not be a fixed size or a fixed number of molecules because it depends on the amount energy received from the sun when it was produced, and if the DNA can vary in size, then the quantity of protein produced by the Ribosome is also a variable and so the amount of new cell growth is variable.

We can see that these variables are able to be changed during particular periods of plant life. When DNA is copied during the process of making seeds, wavelength energy from the sun over a long period of time can cause the new copied DNA to have extra molecules or different molecules, and thus adjust its energy level and quantity to become more capable of producing growth. Thus the specific DNA becomes more 'dominant' in the overall structure of the plant.

Likewise, less wavelength energy means less dominance and less growth.

So during the summer and prior to producing seeds, a plant is actually capable of adjusting its genes to form a better match with the climate and this in turn will cause the plant to slightly change its shape or appearance. I think of this as a form of 'directed mutation', or, as I prefer, 'gene adjustment'.

Seeds can be scattered for miles and even cross continents, but such changes in the local climate will very slowly change the DNA and gene dominance in every cell of the plant and over a very long period of time the plant will adjust itself to the new climate and become a slightly different plant.

The benefit of this unavoidable and inevitable adjustment is that the plant is always re-adjusting itself every summer to match the wavelengths and period in which it receives energy from the sun, hence maximise the energy carried by electrons in the photosynthesis process of growth and survival. It is entirely probable that the wavelengths that a plant uses in the tropics will be so weak in the north that the plant will be caused to change its normal wavelengths completely to shorter, more energy filled wavelengths in order for photosynthesis to occur. Thus the DNA and the genes will be caused to change their basic wavelength, and gene adjustment (mutation) will have occurred.

Note that this is an inevitable process. There is nothing other than the laws of chemistry causing the plant to grow or change itself to suit the climate it is in.

It is this energy that causes cell division to begin and then for photosynthesis to commence growth. We can also see from this that if a forest fern is planted in direct sunlight, cell division will occur at a fast pace and too fast for the photosynthesis process with its food requirement to keep up and so the fern dies.

If a plant that requires intense sunlight does not receive it then there is an insufficient accumulation of energy to cause the genes to initiate cell division and without cell division photosynthesis will not start and no growth occurs.

Somewhere in between these two extremes a plant may have enough energy to produce some growth but perhaps not to the full height that optimum energy would produce.

I suggested earlier that the first transparent animals in a muddy pond or shallow sea changed their cell type from the sun.

## Cell type changes

Whilst the scientific opinion that random mutation of DNA occurs on the helix is a perfectly good solution, I have a problem with anything in nature being 'random'. Nature does not make mistakes! It is a matter of chemistry and physics where there are clear laws. If mutations were random, legs would form in the wrong place and heart and lung creation may be too late for survival. Also if a wrong file was put into a cell it would not necessarily be used because the right energy may not exist, even if it was in a useful place. So I offer an alternative solution.

Every cell is an individual life. The helix is just a file copy. Life could exist without a helix, as it did when created in a muddy pond. So it is cells that drive progress, not the file copy in the helix, and it is each cell that leads to new cell types for the reasons I discuss in the following section.

Wavelengths of energy from the sun through the eyes create more neurons in the brain and more DNA in sperm and egg than are needed by the body. Some of them are used for instinct and memory to pass on to the next generation to aid survival but others are simply passed on because they are on the helix.

Any cell that is in trouble has this pool of DNA available and can select them during the copying process, probably choosing DNA that contains more energy, and energy change is possibly the reason that DNA or genes jump from one position on the helix to the new position where it is required.

The new cell type now produces protein that matches its new energy code and this travels via the blood to the brain where neurons fire electrical energy that matches the code down the nerves to the cell so enabling it to grow. But the same code of energy is passed to the reproduction system where it changes the helix to match that of the new cell type. Thus the next generation is now more fit for purpose and the animal will survive.

There is no doubt that copying errors do occur in the helix of chromosomes, but they are not the cause of evolution.

I believe the causes of gene mutation and new cell types are normal evolutionary mechanisms; cell multiplication in good years, cell interdependence, cell damage, cell needs for survival and muscle use.

**Cell interdependence.**
Every cell in the human body or plant is an individual life form that is fighting for survival. It needs oxygen, chemicals and electrical energy and as the first pond life grew fatter into a worm the innermost cells would have lost their direct access to their needs and instead would force neighbouring cells that still had access to pass them across. Thus some cells became 'supply routes' and these are veins and nerves. So a cell that is dependent on a neighbour will cause the neighbour to change its role and cell type by dragging electrical energy through it, causing damage and forcing it to change its type, perhaps using stem cells, to prevent further damage. One could describe it as the first step in creating an immune and repair service.

A plant requires the same cell interdependence. A leaf requires a branch to hold it up to the sun to gain essential energy and also to provide nourishment, and the branch requires the leaf to provide the results of photosynthesis. In both cases some cells must change their role to form channels, or growth will quickly lead to death.

The mechanism for cell type changes may be similar to a virus which can exchange DNA with a neighbouring cell so that both cells become a different type. All the cells in the body or plant will work together to keep each other alive.

For every life form, every day is a battle for survival. Both plants and animals try to survive. It is a case of whoever moves a leg first wins, not whoever changes a gene first gets a leg. Both are discussed further in the next chapters.

**Muscle use**

Animal needs are such as the need to chew food, hold onto things, cover the eyes from dust, scratch things and avoid things that are causing pain. Need causes attempted use. Attempted use drives an existing cell type to grow to meet the need.

The process that I believe occurred is that every cell directly connected to a nerve has an automatic feedback. Energy passes down a nerve to produce growth, but the end of each nerve is covered with the protein gate that ensures only the correct energy can reach the cell. If pressure is applied to that protein, it will release energy back up the nerve and into the brain, and that is how pain works. But all DNA and every cell type that was produced by the sun has this 'pain' capability and so if the animal has discomfort, the brain knows exactly where to apply effort to achieve an objective, such as to move away from the discomfort. The neurons hold a store of energy captured from the sun through the eyes and so the animal will force the cell's pain mix of energy back down the nerve to the chosen location, identified by feedback.

Applying this energy may only cause a brief jerk, but the animal will keep trying. The effect of this use is to pass this new mixed levels of energy to the reproduction system where it adds molecules to the rungs of DNA in exactly the same location as the existing cell that provided feedback. So the purpose of a gene and cell changes and this is passed on to the offspring. The sun produced the cell type, but use has given it a new purpose. It is a new muscle. And the new energy becomes the rung of DNA that decides future growth from its 'use'.

Note that because the nerve ending has a protein gate, the muscle can only be fired if all the levels of energy for each rung of DNA in the cell are available and sent to allow the energy from the motor neuron to be allowed through. Thus all the cell's DNA mix of energies from a number of neuron's stores of wavelength energy must be satisfied before a motor neuron can trigger a muscle.

One can see that some of this process is automatic and leads to further needs and changes in purpose. If the animal does not have enough energy to run there will be discomfort in the blood flow or air supply and the neurons will automatically send the correct mix of energy to the problem area to create grow a heart and a lung. The same would be true for intestines and so muscles are not necessarily created by the brain, they are created to solve a problem within the body, automatically by brain neurons. And the reason for all of it is that every cell has an automatic feedback mechanism.

**Damage.**

This applies to both animals and plants and I will cover it in the following chapters, but for examples, if a hurricane blows a leaf into shreds, but the plant survives, the changed cell structure of DNA and genes is passed to the seeds for the next generation, because the plant survived. If an animal keeps tearing a joint apart when a muscle is used, pain is caused, so triggering a neuron to force new energy down the nerve to that pain location. The effect is to group the damaged cells into a single new type containing all the DNA of all the cells, so allowing many more chemicals to become involved in finding a solution to stop it tearing.

**Pain cells**

Since all of the above requires some cells to be different in the very first place, and in my opinion some of this happened when the animal was pond life with a transparent body that allowed the sun to pass energy directly to every cell. This stopped when nerves grew that enabled the pond life to grow fat, but the effect was to allow the sun to change DNA when each cell divided. There was no helix of chromosomes at this time to provide control. Later, the feedback system ensured that the change is recorded on the helix for the offspring. It is exactly the same process that produced the pond life originally and allowed it to grow into a worm.

There are bound to be errors in replication and nature has devised a mechanism to try to correct them, and perhaps errors do account for decorative changes such as exotic colouring of birds, but DNA is energy. It cannot be destroyed or dissipated in the short period of replication.

# Cell 'teamwork' and the immune system.

One can see that the adhesion between cells means that a level of teamwork is taking place. If the energy or molecule attraction between the cells stops completely, the immune system will detect a wound has occurred and will produce stem cells to fill the gap and restore adhesion. By matching all the DNA of surrounding cells exactly the wound can heal with no scar.

This energy may well flow right through every cell in the body. Think of acupuncture and ancient Chinese medicine where if the flow of energy is weak, they would conclude that an organ such as a liver, is not functioning properly because its cells have not been replaced, or it has an infection.

# The process of feedback to produce dominant genes.

The concept of gene dominance in which a dominant gene is chosen from the mixing of male and female chromosomes during fertilisation is now well accepted and proven by Mendel's experiments with tall and short garden peas. There is said to be no benefit to the life-form in dominance, it is just a relative difference, but the real question is what is the process that leads to a gene becoming dominant and why does the fertilisation process select a tall pea over a short pea? Surely such a process must have a benefit? There obviously is no 'thinking' in the process of selection so one must assume that the dominant gene just has more energy. That to me implies it has collected more energy from the sun and will grow more. It implies that it is good.

Scientists will say that random mutation produces some favourable and some unfavourable changes and that it is the favourable changes that are dominant and carried forward. But there is no logic in the suggestion that random mutation is important because there is no process – it is random. The whole concept of evolution is a process based on taking the good things forward for the next generation. It is not about luck and the toss of a coin. Furthermore, nature cannot move DNA from one place on a chromosome to another (ie 'jumping genes') without something to cause it, such as an energy change. Changing the position of one DNA could result in a chain reaction affecting many genes that would appear to be a random process but it is not, because something caused the original DNA change.

In order to build on the present to assist the next generation there has to be periodic feedback of favourable things to seeds, where gene adjustments can be recorded and passed on, and this is the real process of mutation.

*So the result of this feedback is that thousands of years ago, the purpose and, in cases of damage, the genes in sperm and egg were not the same as those in the cells of the parents.* Because the human body is now well designed for its purpose and we live our lives always in approximately the same location, most of these changes do not occur. Cancer is one exception where a parent can cause a cancer gene to be created and then pass it on to offspring.

# How did animals get bones?

The field produced by the electricity flowing down the nerve would cause some of the cell types around the nerve to change type again and form into bone type, but also I think damage may be involved.

Clearly the neurons in the brain are all connected together so if the animal tries very hard to move a cell (such as an early fish fin in a worm) and the surrounding cells break or have no strength to assist the muscle, the neurons will seek a solution. But the obvious, and probably the only solution, is to use two or three or four cells together to provide strength with the addition of extra chemicals such as carbon and calcium.

By that I mean several cells will merge into one new cell of a completely new structure, and its gene will contain all of the DNA of all of the cells that merged into one cell. So now there is an entirely new mix of energies which, taken together, gives the new cell the opportunity to use new chemicals from the blood.

One can see that the rule of 'survival of the fittest' must also apply as birds legs shrivel to almost pure bone to reduce weight, and that must be the result of trial and error by neurons leading to survival. So that process, which is only my theory, is like 'mutation, but it is led by neurons to achieve a specific purpose, and will continue until that purpose is achieved.

The neurons are assisted because if cell damage occurs when using a new muscle the pain that is fed back to neurons will identify the problem and where it is occurring. The neurons will then take action to remove the pain and make the cell stronger, as suggested above. I believe such is the case for finger nails and horses hoofs. If the ends of the limbs cannot take the use of the limb without getting damaged, the neurons must fix the problem and make the cells stronger.

Apparently we still have genes on our helix from our early ancestors even though they are never used and the cell they produced have disappeared. So genes never die, they just become dormant and no energy is applied to turn them on because man has no use for the cells. e.g. the tail.

## A summary of the argument that cells need sunlight.

- Serotonin and Dopamine prove that sunlight entering the eyes is a requirement. Why would nature not use all wavelengths of the sun?
- Why would animals have developed eyes if it were not for a more basic reason than sight?
- Sight has a value only after muscles have evolved yet dopamine used the eyes to fire motor neurons before sight had evolved.
- How would peripheral nerves have evolved if it were not a basic necessity to pass energy to inner cells?
- How can cells evolve to become a different type if not using the sun's energy as the driving force?
- Where else could electrical energy come from to trigger muscles?
- Muscles require this energy as a trigger signal, thus their cells do not divide. What other reason could cause this lack of growth?
- Why else do viruses thrive in the sun but have to change their DNA to survive in the darkness inside a human body?

68

- Chemistry and physics conform to a set of laws. How can nature have evolved a process that required random mutation activity?
- UV sunlight changes genes and causes cancer.
- Why would plants make a big effort to place seeds into direct sunlight if it were not to ensure future growth matched the availability of sunlight.
- If neither sunlight nor muscle use could cause gene mutation, would the random fusion of two chromosome in apes really have caused their change to an 'upright' lifestyle? Surely it was the other way round.

## The helix of chromosomes

This is not a critical subject in the evolution of man, but I will make a suggestion on how it may work.

The sides of the helix are made of protein. Protein either allows, or inhibits the flow of electrical energy and I suggest earlier that the protein on the sides of the helix are simply energy carriers capable of carrying every level of energy used in the body. One side would be the inlet, the other the return or 'ground', so each unique level of energy flows down one side until it meets a gene where the rungs of DNA it contains exactly matches the energy. The energy enters the DNA and adds molecules as in 'use'.

My guess is that the genes on the helix are structured in the priority of muscle use, so that those most used for survival are copied first. When apes began walking instead of swinging on branches, legs became more used than arms, and the whole set of leg genes jumped up the helix. Those things never used, such as the tail, are at the bottom and are so weak that eventually they do not get copied.

I believe DNA is not simply the basis of cell type, nor does it define the 'purpose' of a cell. Its main role is to provide its 'wavelength' as an energy match for the chemical molecules that must be used to build the cell. All the energy of DNA in a gene adds up to the energy of its unique protein, and from this the gene is able to control the growth of all of the DNA it contains.

The rungs of each DNA are made of four molecules (nucleotides) and their names are shortened to A.C.G and T. The order or sequence of these four molecules varies, so there are 16 possible sequences. If each molecule had the same 'allowable orbits' as each molecule of protein, then one could see an energy relationship between DNA and protein. Proteins are a combination of Amino Acids of which there are 22 so perhaps 22 wavelengths are used but only 16 form DNA possibly because nucleotides cannot achieve a combination that matches the energy of the other 6 wavelengths. If a rung of DNA is one wavelength of the sun, only 16 wavelength are actually used, and that sounds unlikely, but could be true.

If you now assume that there are always at least four rungs of DNA to make a gene, then there are 16 x 15 x 14 x 13 possible combinations of the four rungs of DNA in a single gene. That would make 44,000 different types of protein, and so the theory could be correct. There is no reason why ten muscles could not use the same wavelength to fire because a muscle cannot fire unless all the muscle cell's DNA energy levels are first satisfied, and these are different for each muscle.

The spider's web shows that an image can be stored as DNA. If each rung is a pixel for an image there will be hundreds of rungs used in an image. Perhaps this is what is called junk DNA? However it proves the link between DNA, the brain and the reproduction system.

But it will take years to understand the helix, and it is not really significant in this book.

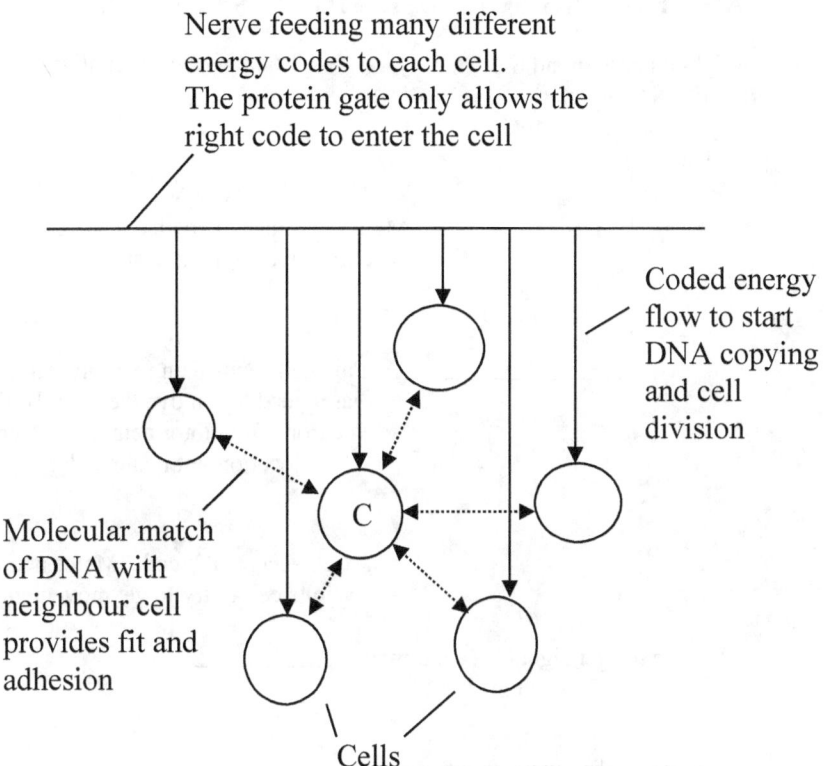

Nerve feeding many different energy codes to each cell.
The protein gate only allows the right code to enter the cell

Coded energy flow to start DNA copying and cell division

Molecular match of DNA with neighbour cell provides fit and adhesion

C

Cells

## Fig 1. ENERGY FLOW FROM NERVES TO CELLS.

Cells of different types must be able to bond together so cell C must contain some of the same DNA as each of its neighbour cells.

A sequence of unique packets of energy pass from the brain through the nerves. A unique protein forms a gate on each cell to ensure that only the correct energy code for its cell type - its mix of DNA- can enter the cell from the nerve.

When the energy that bonds a cell to its neighbour becomes weak, cell division begins. The new cell, with new energy, restores the bond with its neighbour cells.

If cell C were a stem cell created to heal a wound, energy codes from surrounding cells would be used to select its cell type from its pool of all DNA to form a perfect join with surrounding cells.

## Fig 2   MUSCLE CREATION
## AND HOW MUSCLE USE ADJUSTS ITS GENE.

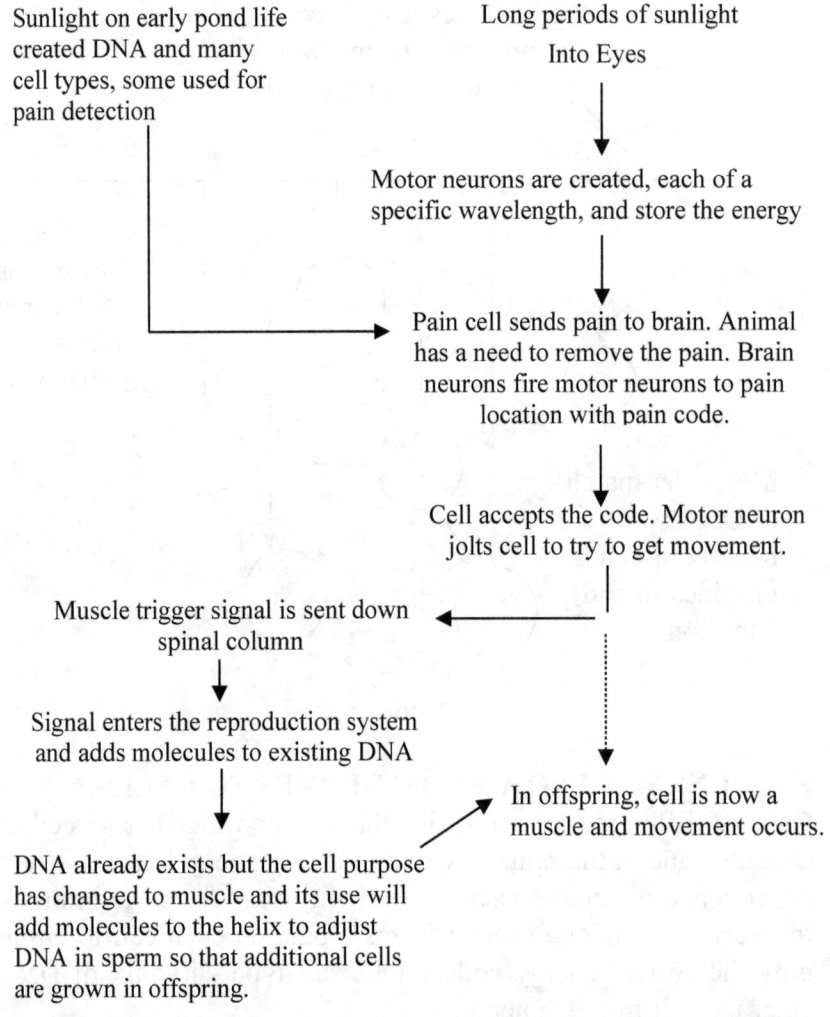

Sunlight on early pond life
created DNA and many
cell types, some used for
pain detection

Long periods of sunlight
Into Eyes

Motor neurons are created, each of a
specific wavelength, and store the energy

Pain cell sends pain to brain. Animal
has a need to remove the pain. Brain
neurons fire motor neurons to pain
location with pain code.

Cell accepts the code. Motor neuron
jolts cell to try to get movement.

Muscle trigger signal is sent down
spinal column

Signal enters the reproduction system
and adds molecules to existing DNA

In offspring, cell is now a
muscle and movement occurs.

DNA already exists but the cell purpose
has changed to muscle and its use will
add molecules to the helix to adjust
DNA in sperm so that additional cells
are grown in offspring.

---

Pain 'cell types' were produced in early life forms by the sun.  A pain energy code forced back down a nerve to the pain location will also go to the reproduction system, eventually changing the purpose of a pain cell into a muscle. The more the muscle is worked, the more energy is passed to the reproduction organs so that the gene is adjusted to produce more muscle growth in the next generation. The entire shape of an animal is set by its muscles.
Neurons are created in the same way but using the sun's energy through the eyes.

## Chapter Seven

### THE RAPID ADAPTATION OF PLANT LIFE TO THE CLIMATE

I will briefly suggest in this chapter how plants have used the energy of the sun to optimise their DNA, and how they have adjusted their genes in reaction to damage caused during growth, in order to optimise their chance of survival. Both are in my opinion inevitable automatic and natural processes that require no intelligence. It may then become clearer how animals have taken that process and improved on it so as to evolve faster and produce the intelligent life of 'man'.

The Jodrell Laboratory at Kew has recently found that plants can adapt to changes in the climate in just 20 years and their research indicates that the plant changes the way in which genes are turned on and off by chemical modifications. I am sure that this is correct but it seems to suggest a temporary solution rather than one that allows a plant to make a permanent and fundamental change in its DNA to adapt to its new location where seeds have blown. My view is that because a plant has the ability to become a life form in its very first location then it must also have the ability to change itself and become a slightly different life form to suit a different climate.

As discussed earlier I believe changes in the sun's energy due to a new location will constantly change genes and their priority, so changing cell type and their ability to grow.

The type of cell that is grown in a plant will depend on the gene that is in control in the nucleus of the cell. The process is that this gene 'asks for' the correct mix of chemicals and it does this via communicating chemicals called ribosome. So the gene asks for the specific mix of unique protein that is required to produce the type of cell it wants.

A gene is a combination of a group of DNA, and since DNA is essentially just a mix of chemicals that has been produced from the energy received from a specific level of energy of the sun, then a gene is a combination of such a range of energy levels. This mixture of DNA molecules has been locked into the gene so that it cannot be undone (except by gradual replacement in the seeds) and when the sun shines, and photosynthesis starts the growth process, the cell can only ask for a specific mixture of chemicals, i.e. a specific type of cell.

A gene is 'switched on' to cause cell division only when the required quantity of energy of the correct mix of wavelengths for the DNA is reached, thus time is required to achieve it.

Every cell contains every gene and so any cell could become any type, but because an exact level of energy in a chemical molecule is required to turn the gene on, only one gene has control. All the others are dormant.

# Gene relationships.

A single gene contains many strands of DNA and my conclusion is that one strand is the 'lead DNA" i.e it is the rung that causes growth, and the other strands I call 'neighbour DNA', because they are identical to the lead DNA of surrounding cells, and that is how a cell is able to 'fit' with and glue to its surrounding cells using a balance of identical energy. This DNA arrangement must be so because all the cells were once of the same type until the sun adjusted some DNA in one cell and this became the new lead DNA.

But the signal down the nerve to permit cell growth must contain the required energy level for all the DNA rungs of the gene within a cell. So some of the surrounding cell's DNA is common to all the genes in a structure such as a leaf. The DNA in a plant where there is no nerve, gets its energy from the sun via a route through other cells because it is the DNA that governs amount of growth of all the cells, or the size of the leaf. (And you don't see fat leaves. The route is either as short as possible, or as long as is possible to conduct energy).

Now consider that for any growth to occur a gene in a cell the lead DNA must receive a quantity of energy from the sun at its designated wavelength, then it is able to start cell division and growth, but the neighbour DNA in the cell, which is common to surrounding cells, must also have their correct level of energy or the gene will produce the wrong type of cell.

I have partially discussed the role of the sun earlier, but, in more detail, if the seeds are blown north, two cases are possible,

If the intensity of the sun and the resulting temperature of seed pods is just a little less than before, the molecules that make up DNA will be able to adjust slightly and still remain within the original gene. But now there is a small 'gene variation', and the effect will be to slightly change the shape of the structure of cells that the gene produces because the energy being used in the growth of each cell is different, and the cells are a slightly different type. e.g. leaves of a slightly different shape.

If the intensity of the sun is substantially different, the original DNA molecules can no longer be supported, but there will be wavelengths that are strong enough to produce a new type of DNA molecule. Thus sun-caused mutation will cause the DNA to change, so producing an entirely new gene and cell type and the plant will change its appearance.

These changes will take place over several years of producing seed pods and in most cases it will be the 'gene variation' process that takes place. The actual ability for a plant to produce an entirely new gene quickly would be rather low and a risk to further growth.

So my preferred assumption is that small 'gene adjustments' occur in seed

74

pods each year as seeds become blown further and further north, until the design of the plant changes quite significantly. In fact it is entirely likely that both processes occur at the same time and that some genes become dormant, others are changed by sun-caused mutation and some simply vary slightly, so that all sorts of combinations of plant structures occur due to the changing climate – and it is simply evolution.

New types of plants grow and their shape will be determined by the mix of energy levels of each gene under the new top gene. Some will grow more than before, others will grow less so the structure has a new shape.

A new shape of leaf (caused by the sun) is not necessarily a more suitable design or structure than the original, it simply grows in that shape because that is how each cell has responded to the change in the source wavelength. Photosynthesis will succeed in the new location and the leaf will thrive because the control gene has reset the operating wavelength for the entire leaf. Whether a plant type survives or dies depends on many other factors besides the climate, so only the best shapes and structure will survive. I.e. The fittest, but damage has an important role in the mutation of plant genes and this overrules any changes caused by the sun.

Root genes of course behave in the same way because their genes are in the seed pods where they become adjusted to the climate.

## The process of adaptation – 'Reverse Engineering'.

My opinion is that gene mutations are a completely natural process caused by changes in the environment of the life-form. These changes in energy cause DNA to move but it is up to the academics to show how energy changes cause changes to DNA. It is just a matter of joining up the dots between the use of Transposons, DNA movement and energy, then collecting a Nobel Prize.

Plants would seem to be the perfect example of errors in copying DNA producing favourable and unfavourable changes to the plant, and only the favourable changes survive, so that plants are perfectly designed to do things such as catch rain, deter animals with thorns and allow the wind to blow through slits in the leaves, but I disagree. Errors in copying are not the reason a plant improves its design!

I will suggest what is really happening and I call it 'Reverse Engineering', I.e. A plant will grow in the following year with DNA adjustments that reflect the end of season design that survived the current year.

All plants respond to the four seasons. They grow in spring and summer and produce seeds in the autumn. Some seeds are in the root, such as potatoes and daffodils, and so there is a mechanism for passing DNA changes through the plant, and I discussed above that the sun is one of the energies for changing DNA.

Where a plant has not developed a flower, the leaves perform the function of monitoring the sun and changing the DNA to suit.

But plants have only embryonic growth. They do not repair damaged leaves and instead they grow completely new ones. They have no immune system and cannot produce stem cells. So the growth is governed by the genes at the start of embryonic growth I.e. the seeds.

Seeds are produced by a gene copying process once a year and this requires energy. The energy from the sun that produced cell division and growth through the summer is now diverted to cell division and copying in seeds, so growth stops. But 'copying' is basically the assembly of molecules to match the helix in an existing cell, and I suggest that the molecules are influenced in that assembly by molecules sent by all cells in the plant, including those that have evolved during the year. The 'current' DNA in the plant at the end of that season has to influence the energy passed to the seeds because DNA is like an energy filter; it will only allow energy to pass that is exactly the same as that which produced it. So where DNA has changed by the sun or by damage, a different level of energy, or a different relationship between DNA energy, is sent to the seeds and this influences the genes in the seeds. One could argue that the plant is 'making' DNA and their relationship in genes, rather than 'copying' it.

The daffodil is an example where the entire plant shrinks down as it creates a bulb and all the knowledge, or design changes gained above ground feeds into the bulb.

Now consider the case of a hurricane that rips a plant's leaves to shreds. Some branches will break completely and their data is lost, but on the branches where the leaves were split, the branches did not break. By splitting the leaves the DNA relationship with neighbouring cells is lost but the leaf does not die. Now in the autumn when seeds are produced, all the information on DNA in the split leaf, which was the only one whose branch did not break is passed to the seeds and is included in their DNA.

So just as the changes produced by the sun are passed to the seeds, so are the changes produced by damage. The leaf sends it current DNA to the seeds and that is the DNA of the split leaf that survived. So next year the seeds will produce a plant with slits in the leaves. Obviously such changes occur very slowly over many years but eventually the leaves are a design that will survive hurricanes.

The same argument can be applied to leaves that catch rain water. In a drought the leaves dry up and curl. Those curled leaves that survive pass their design to the seeds and future leaves have a curl to catch rainwater – because they survived.

Plants with thorns such as the rose, were originally climbers where leaf stems helped them climb up to where the sun is strong and so survive. But some plants would have spread to open land where climbing was impossible and unnecessary. The plants that now survived were those that had a defence against animals eating them. Because the leaves no longer had a role of collecting the sun's energy their DNA would become inadequate. The leaf would disappear over generations, and the stem would grow short; some with pointed ends and some rounded. The leaf stems became protection, but it was the survival of those with DNA of the sharpest leaf stems, or thorns, that were able to survive and feed their DNA energy back to seeds of new plants, whilst those with soft thorns or rounded ends were eaten and did not survive to produce seeds.

So if a weak sun causes the plant to adjust to a stronger wavelength and change its leaf genes, that is what goes into the seeds. If a hurricane splits the leaves, that DNA change is what goes into the seeds. If a drought causes leaves to curl, that is the DNA change that goes into the seeds. A plant adjusts its DNA to produce the design that survived last year!

Sunlight feeds energy to leaf
cells to form molecules of many
different types.

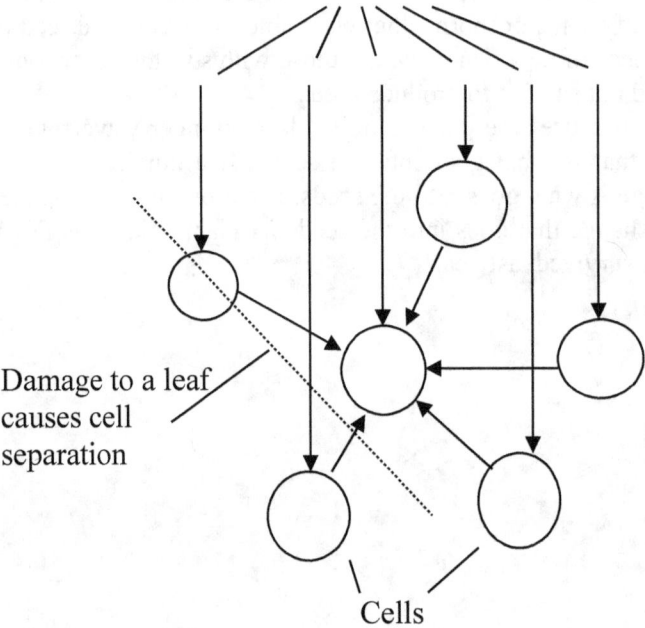

Damage to a leaf
causes cell
separation

Cells

**Fig 3. HOW DAMAGE TO PLANTS MAY CHANGE GENES.**
(Theory)

Plants have no nervous system and instead of electrical energy, the plant
cells produce molecules of unique types from the sun's energy. The
overall effect is the same as shown in a previous illustration (fig 1).

If a leaf becomes torn in a storm the molecule exchange between
neighbour cells cannot occur and the cells will adjust their type.

But if the branch survives the storm the now-adjusted cell types
(combinations of molecules) are fed back into seeds in the autumn. If
enough leaves have this adjustment, over many years, the change will
slowly become permanent and the leaves will have slits like those in a
palm.

# How plants learned to capture energy and adapt to the climate.

The plant Aloe grows about 18 inches high but sends up a stalk 3 feet high with seeds on the top to reach the sun. Another plant similar to Aloe grows wild in the forests of the Bahamas. It grows 18 inches high, but after about 4 years it sends up a stalk 15 feet high! Again seeds are produced at the top, and the top is now right at the top of the forest where the sun can easily reach them. Having used up all its energy in producing the stalk, the plant then dies.

What other reason could there be for flowers to grow and bend towards the sun? Scientist will say that the reason is that each individual cell is attracted to the sun and that may be true, but when I look at the Oleander in my garden all the flowers are on the sunny side of the plant and there are none on the shady side, and when I look at the wild trees in the bush, all the seed pods are at the top of the tree where the sunlight is strongest. So the energy of the sun is necessary for a flower or a seed-pod to be produced and it is this attraction to the sun that has led to evolution.

There is no doubt that during growth, including seeds, when cells divide the genes of the old cell are copied and become the nucleus of the new cell, but as discussed above, I believe that there is an additional process that has led to a much faster rate of evolution, and that is the ability to learn from the current life and pass this on to the new life, cell by cell. This is a process that produces dominant and weak genes, and not simply copied genes.

My belief is that the petals of a flower were originally 'sails' that captured the electromagnetic rays of the sun rather like solar panels by a chemical process where electrons change their quantum level. This collected energy is then passed, probably as a specific energy level chemical molecule, i.e. like protein, directly into the adjacent developing seed, adjusting its gene dominance a little bit more each day during the normal gene copying process.

It could be the reason that flowers need to bend themselves towards the sun. In some plants, that do not have a flower, it is the seed itself that is deliberately produced right at the top of the plant where it can receive the sun's rays directly.

This process is still occurring today but the process has evolved further by mutation causing colours and textures to be added to the petals, with the favourable result of attracting insects and achieving cross-pollination, so that the most dominant genes of two separate plants could be adopted.

# Some plants have neither flowers nor seeds. How do they capture energy?

It is probable that early generations of plants used leaves as petals and had poor adaptability. Nature improved on this by producing petals to do the job faster and allow a greater number of types of genes and cells, so plants developed flowers and many more complex varieties. Then bees came along.

So in early plant life it was probably the sun shining on new leaves that created the right genes. They reproduced by sending out new shoots at ground level, not by seeds. They had very poor adaptation but were a simple life form. Then these new leaves evolved to become temporary devices to detect the sun's wavelengths and these we call petals. And at the same time they produced seeds.

Remember the sun created the genes and therefore it continued to grow in the location where it got its genes, e.g. the shade or bright sun, but it could not easily adapt.

Some sent roots underground and when the root came above the surface the sun turned leaf genes on and it became a new plant. Others send branches down to ground level and when they become covered in earth the leaf gene is turned off and root genes turned on by the much lower energy level below ground. In both cases it is energy that turns genes on – not necessarily direct sunlight.

If a root grows along and then up to the sun, the cells become leaves. It is a matter of energy balance – the amount of energy at a specific wavelength of the sun that is required to build up before the gene can work. The sun has caused a leaf gene to be activated and the level of energy required by the root gene is no longer available.

It may be that because the sun causes mutations these will happen even if the plant does not change latitude. But simply blows into shade. The leaves will grow more and change size – some for the better, some worse etc.

I read that scientists can tell the approximate age of a field by examining the number of gene mutations in the buttercups. Old buttercups will have more mutations. But all the buttercups look similar although the number of petals varies. Surely if mutations were random a buttercup would soon take on a completely different form and would not look like a buttercup? So I prefer to stay with my view that the buttercups simply re-adapt each year to the changing weather at the time of seed 'ripening' in the field, and so each year there are new 'direct mutations'.

A gardener in the Bahamas says that to give roots the best chance to grow, you should plant cuttings at the time of a full moon! There may be some sound scientific reason not yet identified for that result and I can see some logic in it, if the wavelength of moonlight is close to that required by underground roots.

# Chapter Eight

## THE RAPID ADAPTATION OF ANIMAL LIFE TO THEIR ENVIRONMENT

To add to the process of gene dominance that I described in an earlier chapter, when male and female half chromosomes join together to form one new set, the genes that are dominant are selected for the baby. The reason it does so would seem to be nature's way of passing on a function that has been proved to be useful, because such a choice would not seem to be for any other reason. But of course, nature cannot make choices and it is probably the higher number of molecules that causes their selection. Scientists have identified these molecules attached to genes and have called them 'tags'.

Therefore somewhere in the system that copies or creates the genes that form the chromosomes in the sperm and egg that merge into a baby there is a way of producing a gene that has a value that is important and should take preference over the alternative gene. So clearly a system exists that records the usefulness of a function. (Although scientists do not agree that dominance implies good).

As an example, one only has to look at the hoof of a horse to see that the five 'fingers' have become one fat finger while the other four have shrivelled into almost nothing.

In chapter three I discussed that the use of muscles and senses was a clue to evolution based on the spider's web etc., and also that 'instinct' was a clue based on ants etc. I also suggested in the clue from cancer that once a cell's type has been set, the cell is free to grow by itself. So if you now combine 'use' with the independent growth of each cell it can be seen that even the smallest part of a structure can be adjusted by use. A finger nail can become a claw if that is the way the owner has decided to use it.

In chapter six I discussed how I believe a pain cell type is changed to become a muscle, and the same process would produce ankles, toes and toe nails where-ever it needs them.

In chapter seven I discussed that there is a mechanism for the genes in plant seeds to become adjusted to the energy levels being received from the sun in the local climate and weather, and in an earlier chapter I discussed the evolution of nerves and the reproduction system. At this point I just need to confirm that nerves and muscles evolved at the same time because they were inter dependant, and the reproduction organs became located at the start of the tail of early life forms in water.

In animals, the only place where a record of 'use' or a .memory that is 'instinct' needs to be kept for it to be passed on to the next generation is in the seeds in the reproductive organs. These organs are constantly producing new sperm (and possibly new eggs via bone marrow), and it is here that muscle use signals and instinct memory signals must be received and recorded.

I will discuss muscle use first, then instinct.

# Use

How do birds wings become exactly the right size to fly? How does it produce bones? How do bird's beaks have exactly the right strength even though we know from fossils that it has evolved over years to get there? And how do bird's beaks become curved so that they can hook mussels out of their shells more easily? Is this random mutation? No! It all a matter of muscle use.

Wing size is just the fact that the muscle neurons and genes control the entire limb so that if a muscle becomes larger from use in trying to fly, then all the bones, tendons and feathers become larger as well.

Although there is only one trigger signal to a muscles, the muscle was originally just an ordinary cell. If each type of cell is to be able to fit next to its surrounding then the neurons that drive each type of cell, each of a separate wavelength, must put their unique DNA into the gene that controls the entire limb. Thus the gene for a wing muscle must contain all of the DNA for each cell that forms part of the wing. Thus if the muscle of the wing is used all of the DNA in the wing becomes stronger in the reproduction system for the offspring. When the wing is the optimum size, the bird requires less energy to fly so the wing stops getting larger. My theory of muscle use is shown in figure 2 on page 71.

The same argument applies to explain how a bird's beak can become curved. A hooking motion requires more pressure to the top of the beak than the bottom, so the top muscles are used more than the bottom. So the top grows more than the bottom and the beak becomes curved to exactly the shape the bird requires. The bird designed it itself!

The main question is how can muscle use or energy from the sun become a mechanism for creating dominance in genes, or increasing the amount of growth produced by the gene, or even in creating a new gene from the gene of a simple cell?

There are thousands of genes in a single sperm and so there must be thousands of messages being sent down the spinal nerves from muscle use and from the sun in the eyes to assist gene manufacture or adjustment. So that would suggest that genes can only be copied and manufactured to the latest revision level. i.e. it must use the current month's accumulation of 'use' and 'sun' energy, to divide cells and copy and upgrade their existing genes, to produce that month's supply of freshly made new genes in the sperm or egg.

I therefore reach the conclusion that the 'excessive use' signals that the brain and motor neuron send to muscles, or receives from senses, also go to the reproduction system, and these ensure that the genes that are manufactured to pass on to the next generation reflect exactly the needs (genes) that future generations are most likely to require for survival, including any changes in the climate brought about by travel.

It may be that this energy forms separate DNA on the helix ladder which scientists call 'junk DNA' because this DNA does not contain protein-making genes and instead adjust key gene activity up or down, but in my opinion the energy travels along one side of the helix until it finds the gene with that exact energy mix, and then converts the energy into DNA molecules which become attached to existing DNA rungs, so adding molecules and thus a longer period of growth of its cell in the body.

There is one further interesting point and that is our 'reflex action'. If we stand on a nail the pain message is sent to the lower end of the nerve (the axon) coming from a motor neuron which automatically triggers the muscle to move the leg out of the way without any brain feeling or involvement. Because the axon is part of the motor neuron it must contain a store of the correct level of energy for its muscle to be able to do that, but it also shows that all the pain sensor cells (i.e. all cells) are linked to a particular muscle, even though the 'feeling' of pain travels up a separate pathway to the brain.

This connection of all cells to a particular muscle means that the genes that are the blueprint of everything, must record this link and that means that any growth in the offspring's muscles brought about by parental use will affect every cell related to a muscle.

If the spinal cord becomes severed one would expect the reflex action to continue to exist. The fact that it does not suggests that the energy stored in the axon is quickly dispersed and that suggests to me that replenishment of energy by the sun is a requirement of all neurons and particularly motor neurons.

## Life choices.

One can see that animals have a choice in life that is the factor that decides which genes in the body should become dominant from muscle use and I will give you a hypothetical example by considering two families of mice.

Earlier in this book I gave a hypothetical example where two families of mice were washed up onto a desert island where their normal food does not exist. They go their separate ways and one family finds food in the form of fruit and nuts, and security by living up in the trees. It adapts to this behaviour by developing the required muscles during use and becomes a squirrel. The other family finds food in the form of grass and security by hiding in a hole. This family also adapts to this behaviour via use of the necessary muscles and becomes a rabbit. This same approach occurred with apes.

Originally there was just one species of ape but there was a shuffling of genes on the genome where entire chunks of genes swapped places on the helix. The process is known as myosis, and this directly caused the single species to diverge into chimpanzees and humans. These 'jumping genes' are now seen as the main process for producing a new species, but the question is, why did this happen to apes?

Clearly the genome did not decide this all by itself, nor was it a random event because in chemistry and physics random things don't happen. The most likely reason is that some apes decided to live in trees and swing from branch to branch, others decided life was easier staying on the ground, so the gene shuffling was caused because it was the only way the helix could adapt itself to achieve the chosen lifestyles, or 'use'.

'Use' does not just apply to muscles it applies to every sense from touch, smell, pain, sight and hearing etc. Every signal that is sent down the nervous system can be passed on to the offspring, including memories or thoughts such as instinct, (which, by the way, shows that memory is just pure energy attached to electrons that can travel down a nerve). Scientists have even found that 200 genes are associated with academic performance and this again indicates that parental use of the brain is passed down to children to produce higher intelligence.

The theory of use is why every animal seems to be a perfect design for the environment that it is in, as if it had been deliberately designed for the purpose. But it is not a matter of 'creation' or 'intelligent design'. Every animal, including man, created itself to match its circumstances.

One can see that gene changes created by 'use' can create every type of lizard, frog, bird, ant, spider, cat and dog, and so clearly it is one of the main drivers of new species.

## Is the female able to make the same impact?

One can see from this process that dominance of a gene could be produced by the brain in response to the 'excessive use' of muscles. However, it is less easy to see how 'use' and tags produced by females can gain access to eggs as these are produced years in advance and held until needed.

This problem seems to have been confirmed by a study of Red deer by Edinburgh University that showed that gene mutation and 'natural selection' produced good genes for males that were not necessarily good genes for females. Male offspring that were big and strong (i.e had big use tags on sperm DNA) had female offspring that produced fewer fawns, whereas a male that was weak (i.e. had no use tags on DNA) had female offspring that produced many fawns. In the latter case the female gene would have been selected during fertilisation, but with no tags on the females DNA.

Scientists are now focusing on 'Epigenetics'; the study of how genes can be prioritised to be turned on or off while the infant is still in the womb so that the offspring are born with the best combination to match the climate into which they will be born. The start point was that mice born in winter are born with a thick winter coat and therefore a message must be sent to adjust a gene while the infant mouse is still in the womb.

But the study has gone further and has shown that two identical twins with the same genes at birth developed epigenetic differences as they grew up. So clearly the environment or lifestyle has an impact on the way a gene behaves. This seems to me to be a natural process, for example cats grow thicker fur, as winter gets closer. It is unlikely that this is due to the change in temperature as the existing fur will shield the cold, rather it would seem logical that as the days get shorter the amount of energy in the sun each day affects the behaviour of a gene and causes fur to grow thicker and the logical means of detecting this change in daylight is through the eyes. This shows that light entering the eyes has a direct effect on gene behaviour by producing a chemical in the blood just as we know Serotonin and Dopamine are produced in man.

# Instinct.

As discussed above, the record of a very important memory or 'instinct' must be created in the reproduction system from energy passing down a nerve. The memory has been reviewed time after time and in generation after generation, and each time a little more energy is passed through the nerve to formulate a gene and then cause dominance.

How could these 'instinct' genes become unlocked in the next generation? Such genes will be active within the subconscious part of the brain. Their 'memory' becomes a driver of the way the brain responds to situations just as I have suggested in my first book that astrology produces a particular type of personality from the various wavelengths that arrive from the sun, moon and planets, and affects the way a brain responds to a particular situation. Instinct is just a specific molecule mix of DNA that causes other specific genes to turn on and so cause a unique reaction within the cells of the thought process.

And how can a complete picture of a bird's nest and the materials required to build it be stored in a way that can be unlocked and used by later generations? It must be the same process as memory and imagination that I discussed earlier.

We remember things in the form of complete pictures not as individual pixels or signals, therefore all the receptors of the eye are linked into just one memory. But the brain has already made this connection to enable us to interpret vision, so the cells and neurons that make up this vision must be able to store the complete picture and send it as a complete package to sperm genes if it is an important catastrophic picture.

Because the process of handing down memories to create instinct seems to be true, understandable and fairly clear, it tends to confirm that the brain must use the nerves to transfer energy to DNA in the reproduction system where genes are passed to the offspring.

The biggest question perhaps is, how can an image, such as a spider's web, or a memory be passed via genes to the next generation? This is such an important process that I discuss it separately in the next chapter, but at this point it is clear that energy in a nerve can carry such information, convert it into molecules of DNA, where each pixel is a rung of DNA, and place it on the helix.

If we combine the previous chapter on plants with this chapter and say it is unlikely that nature would evolve two different processes, we can say that both probably use 'reverse engineering', the only difference is that plants adjust seasonally in the autumn whereas animals probably adjust in bio-rhythms roughly monthly. So animal muscle use, instinct and image, and plant sun and damage, are all stored locally at cell level and then dumped into DNA and gene production where seeds are made. My guess is that animals store things in neurons, I.e the number of times a muscle is triggered is a memory much like the image of a spider's web.

If an animal survives an attack, a terrible image becomes stored in a group of neurons, with one pixel per neuron, the neurons are changed, and by sending the change down a nerve by reverse engineering, that is how the neurons will be replicated in their DNA on the helix for the next generation.

## The scientific theory.

The scientific theory of Epigenetics is that chemical 'tags' are added to DNA according to the environment in which the baby will be born and that these tags are made when male sperm is produced (which help to decide whether the male or the female chromosomes are selected during fertilisation) and when the baby is developing in the womb.

A tag is perhaps just a molecule of DNA so that perhaps when 1000 tags of the same chemical have been produced, they become complete DNA on the helix.

In this way an image could be produced. If each tag formed one dot or pixel of an image, and the grey scale of each pixel depended on the chemical composition of each identical tag, then 1000 tags could make a reasonable image, and form a group of genes that could be handed down to offspring. Clearly, the image would have to be extremely important or the helix would become full

86

quickly with trivia, hence the image of instinct is limited to catastrophes.

This suggested process has to be reversible so that the image can be 'seen' by offspring, thus the gene must be handed down as part of the embryonic growth process and unwrapped into its 1000 tags in the subconscious part of the brain.

The offspring can now pull this image from the subconscious and see it in the 'mind's eye'. This suggests that the mechanism is exactly the same as 'memory' and that perhaps memory is just the very same tags attached to cells in the brain, but free to move

All of this is similar to the technique used by a simple daffodil to collect rays from the sun, convert them into chemical molecules and transfer them to the place where seed cells are produced, and place the molecules onto the DNA on the helix.

## Moths and camouflage.

One of the most well-known examples of gene mutation is the English peppered moth. A species of moth with light mottled pattern wings rests on Silver Birch tree trunks where its wings give it camouflage. When man produced smog the tree trunks became dirty and the camouflage was lost. But reports say just one moth changed its wings to black so that camouflage was regained and they were protected from preying birds. This was traced to a mutation of a gene in a single moth from which the entire population grew back to its original size. When the atmosphere was cleaned up and the trees became light again, the original light mottled variety came back. Scientists say that this is an example of how mutation can enable survival of the species. I disagree as I do not believe in random mutation.

When I looked at the experiments that were said to show that Einstein's General Relativity and Time Dilation were correct I found that scientists had interpreted the results to prove the answer they wanted. I think the same has happened here.

Why did the gene mutation produce black wings? What happened to the dark grey, dark blue and dark brown moths that 'random mutation' would have produced, were they all eaten? And why did the wing revert back to its original mottled pattern when the atmosphere was cleaned up? That would be a miracle of randomness! The report says that decades earlier a small percentage of moths had black wings and that means exactly the same mutation had occurred once before, so is such an exact coincidence likely?

I saw another study which showed how butterflies change their wing spots to mimic species that taste unpleasant and so avoid being eaten. Their solution was that the wing pattern genes control the pattern by changing the order in which the wing pattern genes appear. Now there must be some similarity between moths and butterflies and so I would suggest that the moth had a black gene and a group of genes that produced the light mottled wing, and it was the controlling gene that mutated so changing which colour gene was in control, so I would prefer to interpret the result that a gene has indeed changed, but it was not a random mutation, it was a direct reaction to the changed environment.

If one wavelength of light entering the eye has changed significantly, it will affect a gene in the offspring. The process is simply absorption of the wavelength of light by the neurons near the eye where it is passed to the controlling wing genes in the reproduction system, to change the offspring. All the moth has to do is look at the tree trunk it is resting on and absorb its colour wavelength, and black is easy to do. It probably actually turns the controlling gene off because black has no energy It repeated the process when the trees were clean and the original wing pattern was restored because the lighter tree trunk was enough to turn the controlling gene back on. Would random mutation get that exactly right? External environmental factors certainly would!

Lizards in my garden change colour from brown to bright green in about thirty seconds and all they do is sit in the sun on green leaves. It may be that fear is the sense that triggers neurons into finding solutions and the act of changing colour is not due to thought, it is simply a nervous reaction because the objective of every brain if to seek contentment.

That is as far I really wish to go in this matter but it seems probable that the helix structure of chromosomes represents the order in which genes must be turned on during embryonic growth, where energy of all levels is sent down one side of the helix so that each step in the ladder is used in sequence so that the cell structure will be correct. Each controlling gene would turn on by receiving the correct energy level in the right quantity and start cell division in sequence from a central gene, and push out further energy, probably as molecules of a certain energy level that would set the cell type of the next growing cell and provide the energy for growth.

And this is why a chicken egg must be kept warm when fertilised. If it were to go cold the signal would fail to produce the required protein from the next gene in the ladder and the whole sequence would stop. (And the hen knows this from its instinct gene!)


So in summary, instead of evolving only according to the strength of the sun and changes caused by damage as is the case in plants, animals became able to evolve by responding to pain or need to create new muscles and then develop these according to the number of times a particular muscle or sense was used, and to avoid situations where an instinct image indicates danger.

The level of 'use' of a muscle by an animal is of course a direct reflection on the circumstances that the animal finds itself in each day, whether it is hunting to catch food or running to escape predators. Thus an animal is able to evolve constantly by strengthening the muscles, or senses of its offspring to better suit them to the world that they will face when they are born.

There is one important difference between plants and animals and that is temperature. An animal has blood held at a specific temperature and this ensures that the entire body, no matter how large, is at that temperature. This temperature control ensures that any specific wavelength from the sun will always create a specific chemical molecule, and equally, any energy that travels down a nerve and meets a different type of molecular structure, such as a gene, will always produce the same result. If the temperature varies by just a small amount, the wrong molecules, or no molecules, are produced and the body starts to go wrong.

This temperature control does not exist in plant life and instead it is water temperature or air temperature that decides whether the correct molecules can be produced. This places a very strict restriction on the plants ability to survive and it is a further reason why plants have difficulty in re-locating to new climates. Animals have the advantage of being able to live almost anywhere.

The above was just an overview of the key processes that forms what I consider to be the miracle of life but the message that I am seeking to convey is that evolution was not caused by random mutations, nor was it caused by some external intelligence or God, but rather it was produced by the plants and animals themselves with a little help from the climate. In all cases the animals had a free choice on what sort of animal they wished to be, and the evolutionary mechanism ensured that the animal rapidly became equipped with the type of body necessary to succeed in the choice.

Clearly adaptation based on the use of muscles cannot bring about basic changes to an animal once the structure of an animal has proved capable of surviving. By this I mean if an animal has survived with four legs, there is no process available for it to start to grow six legs. All it can do is make its four legs fit for the need. Thus the basic structure of animals and insects will have become fixed very early in the evolutionary process.

But reverse engineering offers the possibility of reaching conclusions on heredity, for example; if a parent becomes obese but survives to produce children, the feedback to the replicating genes of the parent will suggest that obesity aids survival, and so the children will have genes that lead to a greater tendency to become obese.

I read a report about wild sheep on a Scottish island that, over a period of 25 years of monitoring, were shown to have become smaller. They now had shorter legs. Scientists claimed that this was due to climate change because the winters were warmer and smaller sheep were now able to survive. But to me it indicates that results of experiments can be interpreted in many different ways, because my explanation is that the longer summers meant that food was more plentiful, so the sheep did not need to walk so far and their muscles and all supporting tissue genes became less dominant, I.e. legs were smaller. If warmer winters meant easier survival, surely the big sheep would also survive?

Likewise Eskimos are believed to have small fat bodies to conserve heat whereas African Masai warriors are tall and thin. My interpretation is that the Masai have to use muscles to run miles hunting for food, whereas the Eskimos just sit and fish.

Whilst muscle use may set the shape and size of an animal it of course cannot set the detail about what happens on the periphery, or skin. For example, Africans have short curly hair that allows air to flow through and perspiration to evaporate and cool the head, whilst northerners have long hair to trap air and keep warm. In both cases the purpose is to control the temperature of the body so that the correct molecules will form, as discussed earlier.

Neurons monitoring temperature (similar to pain) would fire more frequently and produce more protein for growth if they detect skin too cold to create the molecules required, and not fire at all if they detect skin too hot. The neuron process itself would have taken decades to evolve, but once evolved it would seem to produce the same result as muscle use, i.e. the amount of the specific protein in the blood would produce dominant or weaker DNA in the reproduction system so future offspring would have more or less growth of hair according to the past climate.

But again, this is not a process of trial and error through random mutations of DNA, rather it is a precise feedback system that monitors temperature and adjusts growth and gene dominance.

## Man is just a clever Daffodil!

We can see from these two chapters that plant and animal evolution processes are basically the same.

Humans have about 35% of genes that are common to a daffodil. How can that be so when we do not look the same or live in the same way?

The fact that we have the same genes does not mean that we evolved from daffodils; it means that we used the same source of energy and the same raw chemicals in a similar process so that we came up with the same result. It does not even mean that the same DNA produces the same type of cell and the same protein.

My theory is that these common genes are probably the controllers of the process of evolution and I am certain that the same genes will be found in a snail, a mouse and a dolphin not just because they all used the same energy and chemicals, but because without these genes a life form cannot evolve and therefore cannot survive in an ever changing world.

Nature is incapable of making choices, it is just a question of survival, or not. So the process of evolution is self-perpetuating because of survival. It passes valuable information for survival from the parent's environment to their offspring, whether it be a daffodil bulb in the ground or a human baby in the womb.

We have the same seed producing process as plants and we rely on the sun to adjust our genes in the seeds the same as plants and so the basic processes are the same. It is only the creation of nerves that enabled us to have muscles that can draw in air to breath, move, think and grow fat and so we could be called big, mobile plants that are able to think!

So man is just a clever daffodil!

## The creation of new species.

I think it can be seen from these two chapters on adaptation that new species are just the result of changes in the environment and use from life's choices. Such changes may be travel, or climate, or wind, or new predators. My story of the families of mice that changed their environment and became squirrels and rabbits explains everything.

In animals the changes in species are produced mostly by creating muscles from existing unique cell types to aid survival, and then their use to achieve survival is fed into seeds for the offspring, and by the 'jumping genes', which again in my view are caused by 'use'.

In plants the changes in species are produced mostly by changes that were forced upon it by a tough climate, and spreading to new territories, with the changes caused that led to survival being fed back retrospectively into seeds.

It is believed that man became separated from the line of apes because DNA mutations caused two chromosomes to merge into one and this fusion did not occur in other apes and so a new species was born. But I would argue instead that, for some reason, a group of apes moved away from the jungle. There were no bananas and so they ate insects and small animals, and they liked this so much that they began to look for bigger animals. This meant that they had to stand upright, run fast, learn how to throw and begin to think. These requirements were so different that substantial DNA changes occurred and that is what caused the chromosomes to merge.

An interesting question is – what might aliens from another planet look like? If their environment made it necessary for early animals to have six legs, perhaps because of high gravity or a very slippery shore around a pond, would their equivalent to man still have six legs and look like a horse with two arms? Or a man with four arms?

My guess is that such a 'man' would find no use for the extra limbs and their muscles would not be used, so the limbs would disappear and the man would look exactly like ourselves, but with limb strength dependant on the force of gravity on their planet.

# PART THREE.

## ILLNESSES AND CURES.

Chapter Nine

# CANCER

There is so much good information available on the web that this chapter discusses only those aspects of cancer that might be better understood from my theories in earlier chapters on growth, but please be reminded that the following are just my own theories for others to consider. Whilst my argument is based on cases of cancer that I am familiar with, this is hardly proof that they are correct and I do not want to raise false hopes on what is a very serious problem.

## Summary

Cancer is all about energy because that is what DNA is all about, as discussed in previous chapters. To produce a cancer cell there must either be a lot of the wrong level of energy in the body, or some wavelengths of energy are missing.

I suggest that there are two sources of energy that cause cancer: the sun and carcinogens. But there are other factors where energy has a role in cancer, and these are viruses, and inactive nerves. In both cases some of the energy required to turn individual DNA strands 'on', and so turn complete genes 'on', is missing.

The mechanisms that I believe result in cancer are discussed below.

## The prime source of cancerous energy - The sun.

The mechanism is exactly the same as a simple BCC skin cancer, except the high energy is within the body, not outside. I have discussed that energy enters the body through the eyes and is caused to travel down the nerves, where this energy is used both to trigger muscles and to produce cell growth, but there is a third use that I have not discussed previously, and it is this use that produces cancer.

There are a number of cells in the body that are not related to muscles. Their function is to provide liquids that enable parts of the body to carry out their role. These are the reproduction organs, baby support organs, digestive organs, growth stimulant organs etc. Many of these processes use unique nerves, such as the vagus nerve that goes from the brain to the stomach's enteric nervous system that controls the digestion process, and so the 'gates' that control electron entry to a cell are slightly different to other cells in the body.

Whilst tissue cells and cells related to muscles have protein gates that prevent sodium ions from carrying the wrong levels of energy into a cell, the organs that can develop cancers do not have such gates and tend to be hormone controlled. Their role is to provide milk, sperm carrier, digestive juices, thyroid chemicals and other liquids, and in order to do this, there has to be greater freedom by the cells to take in a much wider range of energy wavelengths than normal cells. This is necessary to produce the complex liquids that the body needs to function correctly.

So there are a number of cells in the body that have much less control over the energy from the sun that is available in nerves via the eyes, and it is these that can become cancerous from excessive sunlight.

The process is the same as that which occurred naturally millions of years ago when a single cell amoeba turned into a multi-celled worm, and the changes (which may, or may not be sufficient to occur at cell level) became changes on the helix, forming new genes.

We can now see that just as the sun can cause skin cancer directly by long exposure to a cell, so the sun's energy within a nerve can cause internal cancer if the energy is high and the exposure is long. (Where 'long' means more than one month, but it does not need to be a continual month, it can be short periods of several days, repeated several times, to produce the overall time required to change DNA).

Such energy within the nervous system can cause breast, prostate, testes, intestine, bowel, colon, thyroid, ovarian and other similar cancer. So what decides where the cancer will occur? I think the answer is simply the wavelength that is carrying excessive energy. Each of the cells that I have listed as being at risk, will have unique DNA and it will be the type of DNA that exists in each cell that determines whether it can change by the particular wavelength of high energy that exists.

In these cases, wrong energy in the nerves is not blocked by protein gates, it is controlled to a lesser extent by hormones or other chemicals. When a person reaches middle age the hormones reduce so there is even less control of unwanted energy. Also, cell division and replacement slows down with age so that there is a bigger opportunity for cancer to develop.

I started by suggesting that the sun was the cause, but of course any light energy can contribute to the cancer. So the problem is not just a matter of staying outdoors in bright sunlight for too long, it is also made worse by the bright lights indoors at night, the TV, the Night clubs, and any light that we absorb when what the body is designed to do, is sleep – like a caveman after sunset.

In short, it is prosperity that has led to so much cancer these days. Holidays in sunny countries coupled with bright lights and not going to bed early enough to disperse all the energy collected during the day.

One can also see that stage performers such as Kylie Minogue, Toby Spence (the opera singer) and Roy Castle, who spend many nights in front of bright spotlight, have the potential to develop cancer somewhere. The same is true of those who work night shifts, and those who live in hot countries that are rapidly becoming prosperous.

Brown-eyed Bahamians of African decent get breast cancer at a higher rate than anywhere else in the world, but they never wear sunglasses and now, because of prosperity and affordability of electricity, live the lifestyle of the USA with late nights and TV. Their lifestyle has changed faster than their DNA could cope.

It is understood that smoke in night clubs caused Roy Castle's cancer, but I would suggest that the smoke just changed a gene into a cancer gene and it was the spotlights on stage that really added the energy that caused the tumour cells to grow.

High energy in the nerves is the only way that DNA can be changed in the reproduction system to create a unique cancer gene that can be inherited by the offspring, as in breast cancer. It is the same process that creates and modifies genes during evolution as discussed in earlier chapters, but such a gene cannot be turned on in the usual way because evolution has not given the body a neuron of ultraviolet wavelength, so only another dose of high UV energy can turn the inherited gene 'on' and cause cancer. There are known to be ten different types of breast cancer and I suggest that the variation depends on the specific wavelength of the sun that causes the damage, and the particular DNA that is affected.

Because the body is designed by muscle use, gene copying in the reproduction system is 95% about copying muscle genes and adding 'use' molecules to muscle genes as discussed earlier. It is the remaining 5% of genes that are involved in infant support or liquid production that receive energy that is un-coded and not in 'packets', that are at risk of inherited cancer. There is clearly no protein gate at the reproduction organ, so that all energy is allowed through, including minute changes in energy levels caused by travel to a different latitude, which I have said are beneficial 'caused' mutations. Thus there is a route for UV energy to overwhelm DNA in those organs that are open to any energy, in a way that is not favourable.

In theory, prostate cancer could be caused by a virus feeding off neighbouring cell's energy, but if it were, the cancer would always grow whereas many people seem to have the cancer but it causes no problem. So excessive sunlight may be the initial cause, and the extent of continuing sunlight may be the reason that growth rates vary between people.

If we assume that a helix in a gene in the brain is the routine, on-going method of sending unique codes of energy down all nerves firstly for the purpose of embryonic growth, and secondly, when growth is complete, for sperm production and cell replacement, then it follows that when a cancer protein can be found in the blood it suggests that a gene exists in a cell that does not exist on the helix in the brain's gene. The energy required by a cancer cell cannot be supplied via the helix in the brain and so protein must enter the blood to enable the brain's neurons to release the required code of energy. This process clearly varies as I suggested above that no such neuron exists at the UV level to support breast cancer, and only the sun's energy can produce tumour growth. But it seems to me that once a cancer gene is turned 'on', and held 'on', whether by protein or sunlight, any energy from the sun can produce growth.

Whether a person develops cancer also depends on the particular type of genes they have. Such variant genes have been identified and can indicate whether a person is likely to get a particular cancer, but as so many types of cancer are possible simply from excessive sunlight, it would seem to be an impossible task to identify all the variant genes for all the types of cancer.

It is known that simple regular exercise such as walking can reduce the chance of bowel and breast cancer, and obesity is always linked to cancer, which I suggest is not because the people are overweight, it is because they do not get exercise. It takes a lot of energy to produce a cancerous cell, but once a cell has become cancerous, exercise cannot remove sufficient energy to make any difference to the rate of growth.

Alcohol is also linked to cancers and again I suggest it is not the alcohol that is to blame, it is the fact that the type of people who drink are those that tend to stay up late and party, and do not get exercise. These are also the people who haven't got sufficient vitamin D which is also linked to these cancers.

I believe a polyps is a cancer of a single cell which has very limited connection between the cancer and the rest of the intestine. There is insufficient blood or energy passing through the original cancer cell to permit further growth, so it remains harmless. The ball becomes a serious cancer if a further cell change occurs at the junction between the single original cell and the wall of the intestine.

Prevention of all these cancers – which together probably make up75% of the total - can be achieved simply by wearing sunglasses and getting to bed at reasonable times. Good exercise will help to consume any energy excess that might exist.

I will cover a cure later, but if the cause is prevented, a cure is not required.

## The second source of cancerous energy - Carcinogens.

In this case, the excess energy enters the body through the mouth or nose. We all know that tobacco and asbestos are carcinogens, and prevention is obvious. I am unable to suggest the detailed mechanism that occurs, but it seems to me that, instead of wavelengths of high energy reaching a cell through a nerve, as above, the high energy exists on the orbiting electrons within the chemical particles or molecules, and reaches cells by inhalation. The effect on the DNA would be exactly the same.

But of course it is not a case of a single particle, it is years of particles causing the DNA to change very slowly, and the entire process is slow because the cancer cells must then release their unique protein into the blood to 'ask for' the new energy code to travel down the nerve to allow the cancer cells to multiply.

It is not possible for a carcinogen to produce an inherited gene because the problem is caused directly in the lungs, not via the nerves.

## A secondary cause of cancer - Viruses.

This is known to be the cause of cervix cancer,

I briefly mentioned this in the growth process in chapter six. A virus is probably completely contented living on a pile of garbage and soaking up the energy of the sun that it needs to satisfy its DNA and grow. Then along comes man who disturbs the garbage and somehow gets the virus into his body. This is not what the virus wants because it has lost access to the sun and its genes are turned off. It will die.

So there are only two options. a) To steal some DNA from the host so that it can again receive energy, and b) To change its own DNA to be able to draw energy from human DNA.

In case (a) scientists say it steals human DNA and replaces it with some of its own, so causing cancer. I think there may be a simpler solution, that both the human and the virus DNA are changed in order to achieve 'fit' and share energy. But either way, the virus can now live contented within man, but its cell structure will be different and it has really become a different virus. It has mutated. (Note that this is not 'random' mutation, it is 'caused' mutation).

In case (b) such as bird flu and swine flu that do not cause cancer, the virus must quickly find human DNA that is close enough to its own so that it can adjust its own DNA, just as I described how plants and man adjusts to the sun. (The human DNA is not changed). If it can make the adjustment and draw energy, it will then transfer from man to man easily. It has achieved a minor mutation.

Both types of virus can now travel between humans, live comfortably and multiply, but if some of the sons of the virus get pushed out of the man and into a different man it may find the DNA in the second man is different and does not satisfy its new genes, so another exchange of DNA will be made, or another host DNA will be found, and the virus will become a different one. It mutates again.

One of the problems of all cancers is that cells exchange energy in order to fit together, and this changes their DNA to form a match so that energy can flow freely between cells and hold them together. It is a fundamental life process and it is why metastasis works. This means that a cancer cell will change the surrounding good cells into a cancerous type. It is the normal process (I believe) that adjusts DNA to a changing climate or environment, and similar to the way stem cells set their cell type to fit an injury, but fortunately there is a limit to how many good cells will change

Prevention is by inoculation, as is now being done for cancer of the cervix.

## An inactive nerve

I believe this covers bone cancer. It is a case where there may be enough energy in the body for normal growth, but the energy is not reaching the bone, and so gradually DNA loses its energy and the wrong cell type is produced.

If a nerve has become inactive it is not that the nerve has somehow failed, it is, in my view, that there has been a lack of 'use' to keep the nerve working by pushing energy down it. So if it is inactive it is due to lack of sun, or lack of exercise.

I suggested earlier that bone is a complex mix of cells from multiple fields in the nerve to give it strength and so there will be a mix of energy requirements and possibly more than one nerve. Because the correct mix of energy is not entering the cell as it should, this will eventually cause a change in one of the DNA strands. This change produces a different cell type and prevents the bond with the neighbour cell, so growth never stops.

Prevention is to enjoy an outdoor life with plenty of exercise.

## Melanoma.

Although classed as a skin cancer I believe Melanoma is like any other internal cancer, and is caused by excessive sunlight, but through the eyes, not onto the skin. I believe Melanoma can occur in places where direct sunlight does not reach.

The deeper cancer cell is attached to a nerve and a neuron so that the new energy for growth comes from the eyes, not directly from the sun. Skin cream cannot stop it. The reason melanoma can spread and a simple BCC cannot is probably just its proximity to a blood vessel.

## Benign and Malignant cancers.

From all of the above, it seems to me that a cancer will be benign if its source of energy has stopped. It will be malignant if the energy is still available to cause cell division. A malignant tumour may enter remission if the source of energy is stopped – perhaps as simple as a change in the weather.

It may be that remission that sometimes occurs is simply because the weather has become very cloudy or the person has remained sleeping in bed due to the flu.

# Stress caused cancer.

According to a Dutch survey stress does not cause cancer, and I agree if they mean 'too much work', or 'a nagging wife', but I believe it may be possible for long periods of unsolvable emotional discontent to do so, only because my wife was in such a situation. It is also known that stress slows the immune system and delays the healing process and this time delay in repairing damage could be what allows a cell to become a cancer.

I think the relationship between stress and cancer is an indirect one and it is a matter of whether stress causes lack of sleep or poor sleep.

The prime cause of cancer is too much energy in the nervous system. One purpose of sleep is to allow all that energy to disperse safely away. If stress is allowed to lead to lack of sleep, then it would seem to have a link with cancer.

However it does seem to me that negative thinking people who worry, smoke or drink *because* they do not have peace of mind are more likely to get cancer. Those who have a positive outlook on life and let irritations float harmlessly away, and so get a good night's sleep, will not, whether they smoke or not. I have noticed that the people who get cancer tend to be 'nice' people. They are people who are sensitive and easily hurt, so they are sensitive to not hurting others. So it may be that because they are easily hurt, silly emotional problems bother them immensely and lead to sleepless nights. They are discontented yet cannot find a solution.

I read how an American Jew who had a brain tumour, and a Mexican boy with leukaemia, were both cured by a kiss from Pope John Paul II. Could it be that if someone has faith and believes 100% in the healing power of a person there is a cause to immediately think positively so that all the emotional stress being produced in the brain is instantly stopped by an overwhelming feeling of relief and the excess energy slowly disperses?

I think there is a lot to be said for total relaxation. If every muscle in the body is relaxed there is no energy in the nerves of the body. If the eyes are closed and the brain is switched off, energy will slowly disperse. Some would say that is what sleep is for. (So stop partying).

# The change in human lifestyle.

If we consider that humans evolved from apes that lived in the jungle we can see that the canopy of the jungle would have limited the availability of light from the sun, thus the motor neurons, genes and cells would be of an energy level that would be found in the depths of a jungle, e.g. wavelengths of greens, browns and possibly blues.

Man then left the jungle but was still surrounded by forest and shade. He lived in caves and slept soon after sunset with no light except perhaps a fire.

Eventually man built houses but the Victorians were keen to stay out of the sun using parasols and at night the only light available was oil lamps or gas light.

It was not until the middle of the 20$^{th}$ century that man decided to take holidays in sunny locations and to lie all day in the sun. At the same time electric lights existed in most houses, fluorescent and neon light lit the towns and television kept people up late at night. So it is only in recent times that man has been exposed to very hot sun and bright house lights until late into the night.

Then he began to smoke, drink and invent chemicals to solve problems. Then sex became a recreation rather than a means of showing love, and viruses multiplied. It can be seen that our change in lifestyle is the real cause of cancer.

## Some further clues on cancer.

I read about some research done in Denmark in which adult stem cells that have been left in the laboratory too long will form cancers when transplanted into animals. My emphasis is on their phrase 'too long' because that would seem to imply that cell DNA changes if left too long in daylight. The phrase therefore seems to support my theory on the cause of breast and many other cancers.

We know that radioactivity causes cancer. Radioactivity is alpha rays emitted from the nucleus of uranium and these rays are very high density electromagnetic waves created by the large number of protons in the nucleus of uranium. Basically it is extremely high energy light that we cannot see. The entry point of such waves is likely to be the eyes, even though no light is visible, because the skin would remove most of the energy of these waves immediately. If the skin is not obviously burned by the rays it seems hard to conclude that internal organs can be damaged by entry through the skin. (Although strong doses of similar radiation focused directly at a cancer does of course penetrate and destroy tissue).

So don't look at leaking nuclear power stations! The longer one is exposed to these rays the greater the risk as it is not just the wavelength it is also the amount of energy entering the body, i.e. the duration, that is the problem, and the entire area is likely to remain contaminated for many years.

## Stem cells and the immune system.

One might wonder why a skin graft or organ transplant does not produce a cancer when the DNA of the addition cannot possibly match exactly with the DNA of the

102

surrounding cells?

I think the answer is that stem cells contain every DNA of the person and the energy available from DNA in both sides of a skin graft will automatically find the matching DNA in the stem cells. When this is done, cell type is established and the stem cells will build a bridge between the skin graft and the surrounding cell DNA. Energy will be able to flow across the bridge and growth will continue. If no DNA match is found no healing is possible and the immune system may then kill the skin graft or transplant.

The reason the immune system cannot detect and kill a virus that causes cancer is that the virus has connected itself to a cell's DNA and no bridge is required. The immune system does not detect a problem. The tumour it causes is also connected and growing and so is not detected as a problem. The same is true for all cancer tumours.

A flu virus probably gets energy from blood cells to grow and it is impossible for the immune system to build a bridge from the virus to a fixed cell, or match DNA and so it destroys the virus.

The HIV virus moves around in the blood and probably takes energy from blood cell DNA, then I suspect, eventually from the source of blood cells, so destroying the immune system and becoming AIDS. Special glasses should prevent growth of the HIV virus and AIDS, but will temporarily stop the immune system, so there would be risks, but no greater than doing nothing.

## Cancer that breaks and travels - Metastasis

Eventually pieces of a tumour will break off and travel through the blood and continue to grow in a new location. I have noticed on skin cancer that whilst considerable energy is needed to cause a cancer, once established, only a small amount of energy, and that of any wavelength is sufficient to allow it to grow.

This suggests that the protein wall that prevents wrong energy from entering the cell has broken down, so that once an internal cancer has been formed, and turned on, any energy in a nerve or nearby cell is enough for it to continue to grow. It may be the same process as a virus in that the cancer (which, after all, is just a simple life form trying to survive) will adjust its own DNA to enable it to obtain energy from neighbouring good cells. Thus a piece of a tumour that has become lodged in a new location, is able to grow from the energy available in any of the surrounding good cells that it has become lodged against.

# A possible cure for all cancers

From my analysis, prevention of most cancers is obvious. Wear sunglasses, go to bed early, don't smoke and have only healthy sex. Do this and there is no need to find a cure for cancer.

It is not possible to reverse the change in DNA, therefore there is no real cure for cancer. The only possible actions seem to be to stop the growth process until the tumour exhausts its reserve of energy and dies, or surgery (which does not always stop re-growth, even with chemotherapy etc).

In my view the best cure once a tumour has started is to remove the offending energy. If the energy is removed, the neuron supplying energy for growth must eventually run out. The only source to restore energy to the neurons is from the sun, through the eyes (the process that created neurons with a unique wavelength of energy millions of years ago). If the neuron becomes exhausted then cell division must stop. But such a process has problems!

My first idea was a simple blindfold to cut off the supply of energy from the sun. Whilst this may kill the tumour, it would remove energy from all cells and would eventually lead to a slow death! My second idea was to use special glasses or contact lenses that would block only those wavelengths of the sun that were feeding energy to the tumour, but again, many good cells probably need this energy, so there would be side effects. So my best cure is to draw off the energy from the nerve that is feeding the tumour and, if a way can be found, filter the blood to remove any protein that may be causing the brain to release new energy.

I believe this will work because I have shown that skin cancer cells with minor DNA damage will stop growing and temporarily, and sometimes permanently heal if a plaster is applied to keep the sunlight off, showing that the logic of 'energy starvation' is correct, so if we stop the cancer cell from being recharged with energy from the nerve then the cancer cell division and growth process must stop.

My two suggestions, which are totally new concepts are,

1) To insert an 'earthed' steel needle into the nerve feeding the cancer tumour to drain away the signal and leave it there forever. Ideally, go one stage further and cut the nerve and wrap the end in plastic to stop regrowth and disperse the energy. The problem here is that the 'earth' may draw all the energy from the nerve so that not even good cells will grow and perhaps some kind of plastic isolator around the nerve end may be needed to prevent the nerve growing back until the tumour is dead. It would be better to simply place the steel needle into the tumour itself so that it 'earths' only the unique wavelength of the tumour and allows good cells to grow. The needle should draw energy from all the cells in the tumour. The shaft of the needle should be insulated to protect good cells. If the tumour

104

is large, two or three needles may be required, and the needles must protrude out of the skin by about an inch. It may work without earthing the needle, as I believe such is the case for acupuncture. Charged particles associated with proton radiation, present in all of space and radiating from the atoms in the needle, may pull off the high energy from the needle in the same way as an 'earth'.

2) To inject Botox into the nerve or the tumour. We know that Botox prevents nerve energy impulses from triggering muscles, and that is exactly what is required to stop cancerous cell growth. However this is dangerous stuff and I am unsure of the practicality of doing it safely.

These solutions will work for any type of cancer and there is no need to try to determine the correct wavelength of energy. Cell division should stop and the tumour will shrink and may die.

Inserting the steel needle into the tumour, rather than the nerve will prevent it drawing off every signal in the nerve because the protein gate only allows the 'cancer' signal through, but it must stay there forever. A 'wrong' signal cannot cross from cell to cell. Botox I believe, simply causes a blockage, and so it can be injected into the nerve.

Because the cell stops growing it cannot bond with neighbour cells so there may be a gap, or it may eventually be filled with stem cells, but the surrounding cells will continue to grow normally and should fill and heal the gap.

So it may be possible for all internal cancer tumours, and some virus caused cancers, to be prevented from further growth and eventually slowly shrunk to nothing as old cells die and are not replaced.

Cancers that have broken away and moved are a problem. Such a tumour will continue to grow in its new location by adjusting its DNA so that it can draw energy from good cells. There is also the possibility that, because the cancer cell is out of control and has no specific type, any energy of any wavelength may produce growth, (That seems to be the case with skin cancer) and the break-away tumour gets its energy from good cells. But if energy can be completely prevented from reaching the original cancer early enough, all break away cancers should shrink.

A total cure may only be possible if the cancerous protein is somehow filtered out of the blood.

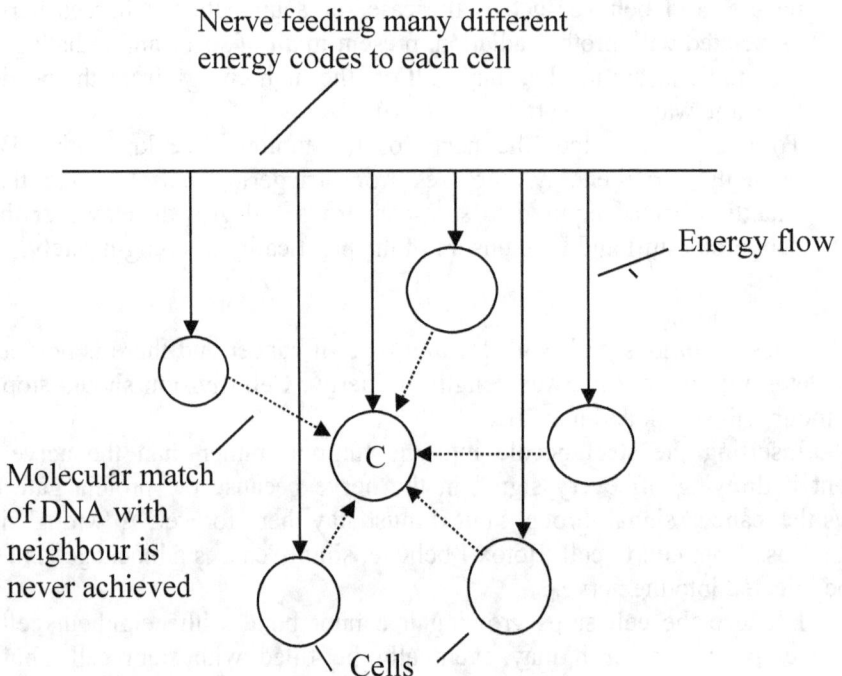

Nerve feeding many different
energy codes to each cell

Energy flow

Molecular match
of DNA with
neighbour is
never achieved

C

Cells

## Fig 4. CANCER IS A CHANGE IN DNA SO THAT CELL GROWTH NEVER STOPS.

DNA can be changed by contact with molecules such as tobacco and asbestos. I also believe that excessively high energy in the main nerve feeding all cells may change DNA in cells that have no muscle to disperse the energy.

Example, if the lead DNA in cell C is changed the cell remains connected to the main nerve and a new signal code will be provided to cause the new DNA to turn on and the cell will grow, but none of the surrounding cells contain this new DNA and so no fit or growth completion can occur and the cell will never stop growing.

If, in the same cell C, the 'surrounding cell' DNA that it contains is damaged, that DNA cannot be turned on and the wrong type of cell is produced that cannot fit with its neighbours. There is no growth completion and again, growth never stops.

When DNA has changed the change cannot be reversed.

# Chapter Ten

## REGROWTH OF DAMAGED SPINAL NERVES

All nerves except those in the spine will regrow and I suspect that the difference is due to the way they were created millions of years ago.

Those that were created by the sun to provide energy to cells, will regrow because the cells down the line must draw energy from those above the break or they will die, and this electric current (because the cell's potential is incorrect) will cause the damaged cells in the break to heal.

The spinal nerves that I suggest were produced by a potential difference, caused by moving in an electromagnetic field, driving electrons through cells to equalise their potential, were primarily used to trigger muscles. The difference in potential is not felt by cells the other side of the break and it is not possible for motor neurons to 'push' the electrons through the break.

I suggested earlier that the amoeba grew a spinal cord because the magnetic waves of gravity added energy to electrons in cell tissue by inducing a voltage so causing the electrons (an electric current) to move towards the head because of the potential difference between the 'earthed' head and the tail. So long bodied animals such as man evolved.

So perhaps the key to growing, or repairing the spinal cord may be to create artificially, a similar potential difference. The problem is that if the spinal cord is severed completely electrons cannot pass through the break and because natural movement is made impossible, no potential difference can be generated to encourage them.

If the nerve is severed completely the person is paralysed therefore no movement is possible even though the motor neurons are working properly. Voltage may be produced but without a circuit it becomes meaningless and useless, and so no re-growth is possible.

The need for a complete circuit or potential difference can be confirmed because if a growing child breaks the spinal cord in the back it will not re-grow even though the child is fully active in growth and has no shortage of thyroid chemical stimulants, nourishment and electric signals from the brain.

Nature extends the spinal cord during the period from baby to adult probably by causing more and more electrons to move down it, so it becomes bigger in the same way as muscle cells become bigger, i.e. a nerve will expand in both length and size to do the job required.

So perhaps the question is; do nerve cells actually have to re-grow when severed, or is it possible to change surrounding cells into nerve cells and then the surrounding cells will multiply to fill the gap? Then, once a tiny reconnection has been achieved, the nerve will expand by itself to carry the electron energy required.

My theory is that if a potential difference is artificially applied to both ends of the severed nerve, this will free up electrons from the good nerve cells to carry the energy level that any cell needs to turn its nerve DNA and gene 'on'. Because there is no direct circuit, the current will affect neighbouring cells so causing these cells to become nerve cells and to divide and grow. This means that new spinal nerve cells are produced by changing the cell 'type' of nearby tissue cells.

## The use of acupuncture to re-grow nerves.

The only requirement of a nerve is that it has the Ion gates that allow electrons to pass through and transport their energy. This means that all spinal nerves are exactly the same structure and that the level of energy that they carry probably has no direct relationship to the level of energy required for nerve cell growth.

If we also assume that the size of a nerve must increase until it is able to carry the volume of signals that are being released by the motor neuron, then it follows that the energy from the motor neuron, whatever its level, must directly cause the size of the nerve to increase. This means that adding any energy causes more free electrons to be released in the nerve. This may mean that the number of molecules must increase.

Thus the size of a nerve must increase when a lot of energy is added and it is not critical what that level of energy is, and this tends to correspond to the theory that the energy of gravity produced the nerve originally even though the level of energy would have varied depending on the speed at which the amoeba wiggled.

If the above is correct then, if there are adequate cells in the nerve area, stem cells may not be essential for surgeons to achieve a repair of a broken nerve and I do not think the origin of the nervous system in the amoeba depended on stem cells. However it does seem to be essential to send a tiny electric current down the nervous system and through the break.

This may be possible by connecting the legs to a machine that moves them slowly from side to side, so reproducing the induced current created by moving through the field of gravity.

Alternatively it may be possible using acupuncture to create a difference in voltage between the upper and lower torso so creating a flow of electrons whose energy may change the type of cells surrounding the break into nerve cells, and as suggested above, any low level of energy can be used, but scientists probably know the 'action potential' required for the ion process to function.

As I said in an early chapter; if a spike is stuck into a jelly and voltage is applied to the spike, nothing happens. You cannot 'push' energy anywhere. But if a second 'earthed' spike is stuck into the opposite side of the jelly and voltage is applied, a current of energy will begin to flow between the two spikes. Energy, in this case voltage, will always try to equalise itself

So if needles are placed near the nerve both above and below the break, the needles below the break are held at ground potential and then a small voltage is applied to the upper needles' a current will try to flow. (It does not matter which way the current flows). The energy will then try to pass through any neighbouring cells and hopefully change their type permanently.

But the problem here is that hundreds of nerves may have been damaged and it is not feasible to connect acupuncture needles to every nerve. So perhaps a better suggestion is just to earth as many of the lower nerves as possible with needles and allow the brain to attempt movement and so release the electron voltage and energy via the motor neurons into the nerves. Use every muscle every day whilst connected to the needles and see what happens.

## Chapter Eleven

### MALE BALDNESS AND OTHER PROBLEMS

# Re-growth of Hair.

This is one of the problems that caused me to start to think through the processes of evolution and growth in the first place, but sorry fellows I have no solution. The following is just a statement of what I believe is the problem.

Scientists say there are many genes that affect hair loss but the most important factor is the level of Dihydrotesterone (DHT) hormone in the scalp. An enzyme (protein) converts testosterone into DHT. This shrinks hair follicles and hair falls out.

I believe hair is slowly disappearing from the human body because we no longer use the muscles. The muscles were intended to raise hair into goose bumps to trap air and keep us warm and also to scare off attackers. We don't have such problems now. The motor neurons are not used and the genes have lost their priority.

The genes have recorded the historic lack of use and a tendency to baldness is now hereditary. Pigs are bald because fat maintains their body temperature. Polar bears have thick hair because of the climate where they live. Cats raise their hair when frightening off aggressors. But we wear clothes, live in warmer climates and don't have much to fear so the genes are useless!

Equally the human brain and skull have become so large that the skin is tight and blood circulation is poor. This affects the availability of enzymes. Testosterone, or its by-product DHT, is the final straw for men. This reacts with the follicles and the few chemicals we need to grow hair. Some men have a different gene that, although weak for the reasons above, is still strong enough and is not affected by testosterone.

Thus gene type and its weakness, DHT and poor blood circulation are the prime causes of male baldness although hair transplants work so poor blood supply is unlikely to be a direct cause. In short, those who lose hair have a specific type of controlling gene and it is of low priority because we haven't used the muscles to raise the priority.

A baby is born with the ability to grow hair and so initially the genes must be adequate, but if the muscles are never used as it grows up, then when energy is redirected in the teens from growth to sperm production and testosterone appears, things go downhill quickly!

There is an old saying that people who go bald at the temples think a lot. People who go bald at the crown are sexy. People who go bald in both places think a lot about sex! There may be some truth in that because thinking may also reduce blood flow to the hair roots.

Ageing also affects the energy available to the genes so slowing cell division and growth, but none of these two factors can be too serious by themselves as women do not have a problem growing hair, so it is clear that the final straw is the specific male gene type and testosterone.

I do have a problem with the concept that testosterone is the only cause of baldness because I cannot believe that nature would produce a chemical in one place that was harmful to the body in another place, unless of course there was no feedback and the body doesn't know that hair is falling out. And why don't other animals such as chimps go bald? No other part of the human body dies unless it is damaged, infected or no longer used.

My theory is that the energy in the nerve that causes cell division is being destroyed by the electromagnetic field in the sunlight simply because it is the roots of the temples and crown that are exposed to the highest sunlight that die first. and the hair then recedes as more roots become exposed. The back and sides where sunlight is weakest do not die even when transplanted to replace bald patches so clearly indirect or weak sunlight has produced different DNA from that on the crown. Clearly the DNA of bald people must be a particular type that produces protein that can be upset when something, probably the hot sun, causes DHT molecules to be produced..

## So is there a solution to baldness?

If we were to stand naked in a cold draft while watching horror movies, or go big game hunting without a rifle, perhaps the muscles around each hair would be stimulated sufficiently and the growth signal would regain its strength at least for our offspring! If we do nothing, then in thousands of years baldness will probably start at birth. Hair on the head will go where the tail went.

Transplants prove baldness is certainly not a matter of poor blood supply. But if my theory is correct, the hairline should not recede if hot sunlight is blocked from the hair roots, so factor 50applied three or four times a day might retain existing hair? Once dead it will not regrow. There is a new device on the market that uses laser light to stimulate growth which the makers claim works. If the right level and quantity of energy can be passed directly onto the hair growth genes, so overcoming that which I believe is destroyed by DHT and sunlight, it may be sufficient to cause cell division and the new cells may have follicles?

I bought a well-known brand of hair growth product from my local pharmacy just to see what it contained. It cost £25 for just one month's supply, and when I read the instructions I was shocked! It said growth would not normally occur for 44 weeks (that's about £250 investment), it may not grow at all and they admitted that they did not know how it worked but might have something to do with increasing the flow of blood! That doesn't sound like a good investment.

In 2007 scientists discovered that in mice, if the skin is severely damaged, requiring substantial new growth, new hair follicles are produced during the new growth and hair can be restored. An embryonic key gene called Wut becomes active. This causes stem cells to migrate to the injury and embryonic growth then produces hair follicles. But I initially wondered about this as surely many accidents and wounds suffered in past wars would have created the level of injuries that this scientific research suggests is a solution, yet there are no 'tales from the past' to confirm that growth is possible.

Many muscles in the body work automatically, such as the heart, the lungs, blinking and I would add goose bumps to this list, but all of these can be manually over-ridden. We can breathe, blink the eyes ourselves and some people can stop their heartbeat or wiggle their noses and ears. If we can do these things then it should be possible to create manual links to motor neurons via brain neurons and learn how to exert the goose bump muscles manually!

Some hope! Give up and accept that bald is beautiful!

## OTHER PROBLEMS.

## Depression and how to check the state of health.

First thing in the morning, either when I have a shower or the first cup of tea, an electric shudder goes down my backbone to my toes, and up into my head. If I do not get this, then either I have a bug coming on, or I need to recharge my batteries.

Animals are electrical machines. We receive electricity in the form of sunlight entering the eyes and we store this in a battery at the top of the spine at the back of the neck. The purpose of the shudder is rather like re-booting a computer. All the nerves are given a shot of energy to make sure they are clear and fully usable, and that includes all the neurons in the brain.

So if I do not get a shudder in the morning, the solution is to get outside into the sunshine and spend ten minutes looking at the blue sky, the white clouds and the green grass. Often that causes a shudder outside, so I know my battery is back up to strength.

Perhaps none of this has any real medical rationale, but it seems to me that if the nerves are not conducting electricity as they should, then cells may not get the energy they must have to divide and grow, and the brain will not be as sharp as it should be. So some cells may die before they should and we become old before our time, and we may feel tired because the brain is not working properly. I suppose it is even possible that some genes will not get turned on so that, over time, they stop working, leading to bigger problems.

113

Walking and other exercise helps to keep the nerves working, and being out in the bright sunlight and fresh air keeps the body fully charged up. My solution is to play golf.

Depression is very similar to that analysis, as it is due to insufficient sunlight leading to a lack of a chemical called Serotonin being produced by a gene. It demonstrates that energy plays a strong role and that a continual supply of new energy is necessary for a stable mind. I suspect that this depression is rather like 'lack of use' in that without this continual supply of energy the brain doesn't get the molecules it needs.

So the answer to achieve contentment is to give up the car and walk in the sunshine (but wear sunglasses if the sun is very bright). If we did this, then the brain achieves a better sense of balance, and problems will just wash over without concern. Instead of thinking negatively you will think positively.

If you are having a nervous breakdown, don't take Valium, walk five miles every day for a week or play golf. If this theory is correct, playing sport in schools is vital. I also think it is possible for anyone to learn to think positively and grow ones 'contented' neural pathways.

Almost every illness these days is being blamed on a variant gene, yet these genes are not always turned on because they require certain environmental factors to exist and so have no effect. Furthermore, according to a report in my newspaper, in spite of all the good work on genes there does not seem to be a single successful new treatment in use to prevent a disease arising from an inherited variant gene.

There are even studies that show babies born in the spring are at a greater risk of developing a disease than autumn babies. Conditions such as asthma, Alzheimer's and MS are all linked to the mother's exposure to sunlight during pregnancy. The reason is said to be possibly due to vitamin D which has a role in regulating when genes are turned on. My suggestion is that the theories in this book are being proved right. The actions of a gene, even in a baby, are affected by sunlight entering the eyes and passing more, (or less) energy into the nerves.

# Obesity

This problem seems to be getting worse. The simple answer is just stop eating! Eat only when you feel hungry and eat nothing between meals No biscuits, chocolate or cupcakes. And get some energetic exercise to burn off the calories.

But even if some people do that, they still have difficulty losing weight. It is as if the genes will not permit it, and that, I believe is one of the problems.

In my opinion, to understand the problem you have to consider the mechanism of life itself, and that is so incredible it causes you to think that there must be a God. Every lifeform is somehow programmed with an objective to survive.

I discussed earlier that everything we do in the few months before conceiving a baby, is stored in the sperm or egg. The DNA keeps a record because of that prime objective of survival. The genes in our reproductive system record what we have been doing because we have survived, thus what we have been doing must be necessary for survival. The process is an absolute miracle of life, and it is true for all life forms from viruses to daffodils, jellyfish and humans.

So if our parents were fat, their genes have recorded that being fat is necessary for survival, and the parents have given their genes to us. So some of us have genes that require us to be fat because that is how the body has recorded the correct weight to survive.

That makes losing weight for some people very difficult because their genes will not permit it.

## Parkinson's disease.

We know that the sun enters the eyes first thing in the morning and produces dopamine, and dopamine is used by the brain to send signals to motor neurons when a muscle is to be used. It is also known that Parkinson's disease occurs if the supply of dopamine falls substantially or if there is a loss of function of the neurotransmitters and response neurons. These things happen due to a particular gene and so there is an inheritance risk.

Dopamine is a store of energy that neurons need in order to send messages, and it represents a specific wavelength of the sun.

Whilst there may be a deficiency in the cells that produce dopamine, that seems to me to be unlikely, and the most likely cause of Parkinson's disease is life style.

If someone works in a dim office doing a brain intensive job for long hours, dopamine is used and is not being replenished fast enough because there is no sunlight.

Equally, sitting all day at a desk means muscles are not being used.

I have suggested that nerves grow big enough to do their job because they are one cell that has changed in order to pass energy to another cell, but if they are not asked to do anything they will shrink.

So after doing this job for many years, nerves going to muscles will shrink from lack of use, and at the same time the dopamine energy is being produced in less and less quantity each year.

So a stage is reached when the nerves are thin and unable to carry clear signals from motor neurons and the neurons cannot provide adequate energy to keep the motor neurons turned on.

The prevention and possible cure is to walk in the sun outdoors at lunchtime every day. If Parkinson's disease is already bad it may take a considerable time to get the system working again. The best cure is to play golf.

The more sun and the more muscle use each day the faster things should return to normal but it may be a more serious problem in the brain cells than my simple solution can put right, and it may be that nerves cannot regrow after a certain age.

If one looks at all the symptoms and apply some logic as well as the theories I have discussed earlier we can begin to see the problem.

The symptoms are,

- Movement disorder.
- Lack of sleep.
- Depression.
- Visual hallucinations.
- Speech difficulty.

These are all related. Movement problems indicate insufficient energy in the brain to drive motor neurons. Lack of sleep suggests that the brain is exhausted and has produced too much adrenalin. The lack of sleep taxes the brain even further and increases the problems. Depression is in my view as discussed earlier, a lack of the right energy in the part of the brain that decides contentment.

These all indicate that the brain has been used too much and has run out of energy and there has been insufficient exercise to remove adrenalin. Relaxation therapy is believed to help and one can see that this is an alternative way of dispersing adrenalin whilst giving the brain a rest.

Scientists say that nerve cells have actually died in the part of the brain that produces dopamine – the chemical that induces movement. That may be so but nerves will re-grow if energy exists in the right place to cause it, so I believe the disease is curable but will take time, and with age, it will be a slow process.

The brain gets its energy directly from the sun and so exercise in the open air is critical. Here is my suggestion to prevent the problem,

1. Fix a strict and low time limit each day on the use of the brain and take frequent breaks for tea or coffee.

2. Have complete breaks at least twice a week in the form of walks in the country where there is blue sky, green trees and grass. Colours are important and avoid noise, advertising signs that you automatically read and traffic.

It is said that there is an inherited faulty gene that contributes to Parkinson's. Perhaps so because if the parents had the same lifestyle as that discussed above to cause Parkinson's then a gene mutation is a possible outcome. But I do not see the genes to be a problem as this would cause Parkinson's to occur at any age.

## Alzheimer's disease.

A set of five genes are now believed to affect the likelihood of developing the disease, but all diseases now seem to have a variant gene involvement and I find that unhelpful as it tells us nothing until it can be shown that these produce a unique and harmful protein. Alzheimer's disease is known to be caused by the build up of a protein called Amyloid. This blocks the blood supply and cells die. Pharmaceutical companies are close to releasing a drug to prevent this but their objective is to get you to take an expensive pill every day for the rest of your life, whereas I would prefer something that prevents the cause. Every protein is produced for a reason and has a use, but if it is not used, then it probably will clog things up.

Whereas Parkinson's disease is over use of the brain and lack of sun and exercise, I believe Alzheimer's disease is partially the reverse. It is too little use of the brain neurons causing the nerves that join them (synapses) to shrink so that memory and control become lost, and too little exercise to force the blood to clear Amyloid plaque build up so the cells die, and there must be an unwanted DNA variant in a gene. My hopeful solution to avoid the disease is to work on this book and jog along the beach once a week. I refuse to blame everything on genes or to cure everything with drugs.

## The strength of faith and homeopathy.

Positive thinking is rather like faith. Faith and the placebo effect seem to be the same thing and scientists have shown that a placebo does have a benefit because if the brain has hope it produces a chemical that can reduce pain.

I believe we have to consider the believed benefits of homeopathy in a completely different way. If we believe that some illnesses are actually caused by the state of mind that we are in, such as feeling lonely, un-loved and un-cared for, then if someone comes along and says I have a complete solution for your illness, he is not actually providing a solution, he is showing that he cares, and that caring immediately removes the mental cause of the illness. The placebo effect is probably exactly the same.

The mind is a very powerful force! If we believe a drug can help it probably will. If we believe a homeopathic mixture of some awful substance such as newts tails boiled in a cauldron or sulphuric acid, but is diluted to such an extent that it is totally harmless and useless, we will feel better just because the bottle has the name of the awful substance on it. "If someone tells me that swallowing boiled newts tails will solve my problem, and I have the nerve to do it, it must do some good"!

Chemistry is not my strongest subject but it seems to me that the chemistry of homeopathy should not be ruled out! The free electrons that surround a molecule mixed with water, will have different energy to those surrounding just a water molecule. This energy must transfer both ways, so that the water gains a different energy, and it is the energy on the electrons that provides the cure. The energy cannot be washed off the water no matter how much it is diluted, but it is only of value if held in a non-conductive container or it will disperse as soon as it meets with a conducting surface.

## Acupuncture

There is considerable doubt whether this form of medicine actually works but I believe it is possible.

A steel needle piercing the skin will draw energy from the cells it touches and from the nearest nerve, because each cell will draw energy from its neighbouring cells including those that touch a nerve. The high energy on electrons in the proteins or other chemicals in the cells and nerve will transfer to the lower energy on the electrons in the steel needle. Very low energy positively charged particles involved in proton radiation, or even moist air itself, will draw the energy off the negative electrons in the needle and carry it away into space. Thus high energy, which is just a complex code of the sun's wavelengths, will be drawn out of the body and disperse into the atmosphere.

If there is pain in the body, removing the energy will reduce the pain because it cannot travel to the brain. If any illness causes an abnormality of energy then that abnormality will be removed by the needles. I believe the ancient Chinese new more about energy in the body than we do today.

By removing energy, such a needle may start cell replacement in all the cells it touches and so stimulate healing. Equally, if it draws energy from a nerve it may prevent cell replacement of any cell attached to that nerve. It is the same process that I proposed may be a temporary solution for cancer. It cannot kill the cancer, but can shrink a tumour to a harmless size.

# PART FOUR.

# CONCLUSIONS

# Chapter Twelve

## TRIVIA

## Which came first, the chicken or the egg?

Some English scientists have decided that the egg came first because the DNA of the egg must be the same as the DNA of the chicken that laid it. But if that were true it would be impossible to have evolution?

Evolution can occur only if the egg has different DNA so that the chicks that emerge are one-step better models in their capability of survival than their chicken parents.

This means that the DNA that is put into the egg by the parents has been changed, enabling the chicks to become better chickens! It is a process of learning that leads to better instinct, and use that leads to better muscles.

So there is no answer to the question of which came first because the chicken and the egg are not the same animal.

## Fibonacci Growth

Fibonacci numbers are a sequence of numbers in which the first two numbers are 0 and 1, and each subsequent number is the sum of the previous two and the results form a natural pattern:- 0, 1, 0+1=1, 1+1=2, 1+2=3, etc. In short, 0,1,1,2,3,5,8. But the interesting point is that many plants form their seeds in such a pattern of increasing seed size so forming a distinctive shape, such as sunflowers, pine cones, pineapples and also snail shells.

Of course nature does not use numbers, they are just man's invention to describe such a shape. It is simply a growth mechanism following a spiral shape in which the outer seeds are more mature and larger than the newly formed seeds at the centre. But the growth is not a matter of 'pairs' as in the sequence, it is a matter of simple addition of the preceding seed number to its own number, so producing growth.

One can guess that there is some molecular movement occurring where an older outer seed receives the completed molecules of its inner newer seed to allow itself to mature. This means that all molecules are being added at the centre and are passing through the structure from the centre seeds, outwards. So an outer seed needs the capability of the inner seeds for it to mature.

A seed that has 3 molecules will receive two new ones from its inner neighbour and become 5. This 5 seed will then collect the 3 molecules from its neighbour which has grown into a 3, and become 8.

Whether these molecules form DNA or genes or the helix structure is too complex to comment. It is probable that a helix is the source of the sequence.

Fibonacci growth is a good example of how energy passes through from one cell to another to enable growth.

## Is there a second life – Reincarnation?

I read in the Miami Herald in January 2005 about Dr. Brian Weiss, a Miami based psychiatrist who has focused his career on past-life regression therapy. That means to take patients back to a so-called previous life through hypnosis in order to correct mental disorders and phobias. Dr. Weiss is an intelligent man educated at Yale, so that his ideas must be considered as valid.

You will have realised by now that my approach to questions such as this is not to look at the success rate of Dr. Weiss, nor question whether the stories of people's past lives are true or imaginary, but rather it is to consider what possible process could exist to make such phenomena possible? I wondered whether my new- found knowledge of the universe could offer any clues as to whether we have all had a previous existence, and if so how could it work?

After some thought I reached the conclusion that there is nothing up there in the universe that could support such claims. A memory system would be required somewhere in our galaxy, and whilst that is not impossible it is pretty close to it!

But then I considered the more obvious solution of genes and 'instinct'. If one considers genes as being tiny microchip memories rather than biological bits, and if one also considers my earlier suggestion that we can influence our genes by using a muscle more, then one can pass on a particularly vivid and terrible event experienced by an ancestor as a 'fear' to protect future generations. Hence phobias such as claustrophobia emerge. The process would exist as a warning system with the intent to protect and prolong the species.

The gene containing this warning would implant the memory into a new-born brain and depending on the astrological personality of the person, the memory would emerge either as a phobia or just as a normal awareness.

The concept I have suggested is not the same as religious beliefs of having a previous life, it is simply a warning system that passes on life- threatening memories. Nor is it a 'soul' from an earlier person, although the memories do come close. It is just 'instinct'.

Such a warning system would have to include the details of the specific environment, or the situation that arose in an earlier person, for it to have any relevance and that may be the past-life that emerges through regression.

If true, it would mean that the person undergoing regression therapy is a descendent of the person in the past-life that they recollect during the regression, and so something that could be loosely described as a 'soul' is past on. Regression is not a process of going back in time it is a process of delving deep into the subconscious part of the brain to find the smallest memory.

In 2008 there was a programme on TV that included a discussion on 'spiritual release'. This is suggested as a mechanism where, under hypnosis, the spirits of other people that are occupying the body can be removed.

I suggest that this is a misunderstanding of the problem. It is not that the person's mind is occupied by spirits, it is that hypnosis is a method of gaining access to, and diluting the relevance of 'instinct' that has been passed on, into the brain, through the generations. There are no 'spirits' in the brain.

But there is more to the feeling of reincarnation than discussed above because there are other strange cases such as 'the boy from Barra', and 'the ladies from Bath'. In both of these cases the reincarnation is about people who died and therefore it is not possible for the 'feeling of death' to be passed on through the genes.

## Astrology revisited.

Scientists do not believe in astrology but I do! I do not mean the daily horoscope or any type of fortune telling that seeks to forecast the future, I mean the basic personality that each person has due to their date of birth.

The basis of my belief is similar to that I have discussed for neurons and motor neurons, where external energy enters the body at birth and sets the neurons at a particular energy level or wavelength, There are two possible processes that could influence the brain at birth and I do not yet know which is correct.

Firstly, all planets, the moon and the sun have gravity because they are made up of atoms. Gravity is produced by positive proton radiation from all these atoms, (read my first book). The level of energy of this radiation depends on the type of atom because atoms with more electrons and protons will radiate significantly more energy. Uranium is an example. This energy can pass through the skull just as gravity passes upwards through the ground. So when a baby emerges from the womb it will be bombarded with these fields either through the skull, or through the eyes, so influencing the brain's neurons.

Secondly, the moon and planets are all rotating in an orbit either around the sun, or around Earth. When atoms are caused to change direction the electrons within the atoms must radiate some energy. This energy is a negative wave that is not attached to a particle. Such a field or wave could also pass through the skull, just as radio waves can pass through buildings, and cause a change to the brain's neurons.

In both cases the energy in the radiated fields will be different for each planet or moon simply because their make-up of atoms (their material) is different. This means that their effect on the brain's neurons will be different. The strength of influence of a planet or moon will obviously depend on the distance from Earth, and this varies throughout the year as they travel in their orbit, moving closer or away from Earth, thus each birth date can produce its own unique personality.

Personality exists in sixteen forms and these come from various combinations of just four variables and therefore one can guess that this is due to various combinations of just four neurons or genes. Only two factors are necessary; a transmitter and a receiver. The transmitters are the moon and planets. The receivers are the chemicals that formed the neurons in the brain.

So astrology is rather like instinct except instinct is a memory that informs a young life what it must do to survive, whereas astrology produces character guidance on how to react when unusual situations arise.

This theory is re-enforced by the well-known effect of the moon. The strong electron radiation from the moon can change the way we react to daily situations by stimulating certain neurons and I suggest that it is the close proximity of a full moon that causes some of us to do lunatic things. The ability of the brain to both send and receive information by one, or both of the methods discussed above is given further support in telepathy and the story of the boy from Barra in the following sections.

## Telepathy.

Telepathy may be possible because the atoms making up the cells in our brain are radiating electromagnetic waves all the time and it may be possible to put messages into these waves in the same way as radio.

It seems probable that any animal that is able to think can have the ability of telepathy. I am thinking particularly of dogs who seem to know immediately when their owner is coming home even if the owner is miles away.

But simple telepathy is not a matter of transmitting pictures or shapes as people have tried to test in a laboratory and failed, it is rather a matter of transmitting emotions or feelings such as anger, fear, love, hunger etc. Nor is it possible to pretend these feelings because the feelings come from the subconscious part of the brain that detects 'contentment'. So telepathy is really an unintentional transmittal of feelings that cannot be prevented because it is an uncontrollable radiation. It is an electrical signal attached to the waves that are radiated by every atom in the brain. It is almost the reverse of astrology.

In fact I think we are almost capable of telepathy already. When someone talks to us we look into their eyes and subconsciously check that their eyes agree with what they are saying.

So that in my view is the reason that a dog knows when its owner is coming home. The owner is happy and the dog detects it. Presumably the dog also recognises the wavelength of its owner so that it does not confuse someone else's happiness with that of its owner.

# The Boy from Barra.

This is a phenomenon that nobody can explain and I have difficulty finding anything in the evolution process that could have produced the effect that seems to have occurred,, but I have an idea.

'The boy from Barra' is a factual story that has caused substantial investigation. The boy, when only old enough to talk, was totally certain that he had lived an earlier life. He could describe his old house, the village in Scotland by the sea, his dog and how his father became run over and died. He described his own death as 'falling down a hole'. His mother and Dr. Jim Tucker of the University of Virginia visited the village that he described and found there to be sufficient accuracy. The boy had described how, from his house, he could see aeroplanes landing on the beach, and there they were!

The only conclusion I can give is 'telepathy at the time of death', which is rather like the concept of 'the spirit leaving the body'.

The brain of the boy who died would have contained thousands of separate memories of his house, his dog and the aeroplanes that landed on the beach. When the boy died and his brain stopped, the bit streams of these memories would no longer be contained within the brain, they would become free to move. Waves produced and radiated away by every atom in the brain (defined in my earlier book) could then collect these bit streams and carry them away through the air.

If at the precise moment when the boy's brain stopped, the eyes of a second new baby boy opened, the bit streams could become lodged in its brain as a flood of memories, and the baby would grow up to assume that it actually saw these things.

The likelihood of this happening is extremely remote but if there were thousands of memories sent by the first boy then it is possible that the second boy could receive some of them.

The same may be the case for 'the ladies from Bath' who all recollect living in the 13th century as 'heretics' who were brutally killed by the pope. The method of death was so horrific that the brainwaves released by telepathy (my view) would have been so strong that babies being born at that very moment of death would pick up the waves and remember them. From then on the 'memory' would have been passed on in the genes to later generations.

That is all just my theory and it is a weak case!

# Awareness.

If some genes can hold memories of 'events' it becomes a little easier to understand what consciousness actually is. It is this 'event' gene implanting key instincts into a new- born brain somewhere in the subconscious. A few years later it is the feedback that a young brain receives when it makes objects move by pushing them. These instincts and feedbacks cause the brain to realise that it is actually a person that can change its surroundings in order to protect itself and become contented without any help from a parent. It is the 'soul' of a person.

Once learned, doing things oneself becomes automatic because it is faster and less trouble than trying to get someone else to help.

Every living thing that can move by itself probably has a feeling of fear because avoidable action is possible, and so they probably have an event gene and consciousness in some form. Things that cannot move, such as a tree, probably do not have fear as it would serve no purpose, and therefore probably do not realise that they exist.

Contentment drives everything and it is probably nothing more complicated than the wrong molecule on a particular neuron in the brain. If we know when we are unhappy; we know we exist.

Because instinct gives us a basis for comparison as life goes on, we have something to refer to, and this in built memory helps us to know that we exist. We become aware that we are a person when learning takes over from instinct. It is when an action by us leads to an event and we realise that we can cause things to happen.

Then, of course, there are the obvious things such as the senses, hearing and language that help us to communicate with, and be aware, of our surroundings.

## Life expectancy – From instinct, choice and intelligence.

Once puberty is reached, the cells are in control of their own replacement and they do it via protein and energy through the nerve, so in theory, if we get adequate sunlight in the eyes to put energy into the brain and cell controlling neurons, our subconscious neurons should continue to trigger cell division and growth forever, but of course that does not happen.

In the chapter on clues to evolution I discussed that a spider builds a web from instinct handed down through generations. It does not really know why it should do it. This concept can be taken a stage further.

Just assume that suddenly one spider has intelligence. It considers the situation and decides that a six inch diameter web will catch bigger and better flies

126

than the three inch web that instinct drives him to build.

The spider builds the six inch web using every muscle in his body to exhaustion, then looks at his finished web with pride. He then produces baby spiders with his wife.

A baby spider grows up and finds that he has bigger muscles than dad, but he is confused. His instinct says build a three inch web, but he has a vague second instinct that perhaps it should be a six inch web, the same as dad builds. He has to make a choice.

If he builds a three inch web he will feel guilty because it's not what dad builds. If he builds a six inch web he knows it will be hard work and is not convinced that he really needs all the big flies that such a web will catch.

So he settles on a four inch web. He has gone against instinct and against the family, and even against the culture of a three inch web. But he had a choice and intelligence.

So what is my point in that little story?

1. We all have a culture that is built up over generations and has become an instinct recorded in our genes. It becomes difficult for an Englishman not to want afternoon tea when he is living in India. It is difficult for an Asian or an African to accept the culture of England when it is so different from their instinct.

2. If we apply intelligence to our ability to choose, we can slowly change our culture and improve our way of life.

3. Evolution has done this automatically by survival. As an example, perhaps those spiders that built six inch webs did not survive because they died of hunger before the web was finished. There is only one best answer to the size of a web and survival produces the answer.

Culture is deep within the genes and we can only change it over a period of several generations, but there is interaction between 'instinct' and 'now', the real world. As intelligent people we are constantly battling for change and improvement, but our instinct holds us back.

Sometimes we cannot tell the difference between a 'visions from instinct' and a 'visions we actually see'. We do not know which is right. We have become two people inside one. We have become schizophrenic, and there is guilt because we no longer know with any confidence which is really us.

These concepts are closely linked to our expectations about the length of our lives. Over many generations we have decided the length of a useful life and this has become an instinct. But there must be a link between instinct and the life controlling genes that decide when cells should divide and grow, and at what time that process should slow down, because they are all in the same place in our brain and all communicate with each other.

# The Secret of a Longer Life?

Scientists know that as we age the process of cell division and repair slows down until organs cease to be repaired and we die. They have identified structures called Telomeres, or Tags, that protect the DNA strands as they divide during cell copying. These Telomeres become shorter with each cell division until there is none left. Then copying causes DNA damage and repairs are incomplete. It would seem that perhaps the Tags are a store of energy that have replaced the brain's helix sequence of energy released into nerves, and the Tags provide the 'permission' energy for replacement. This energy is used at each replacement until there is none left.

But the question is, how are the lengths of the Telomeres established? Clearly this is done at the sperm and egg stage and so cannot be changed after birth. In my opinion, the answer to its length is 'as long as is required to do the job'. There is no external force, or force of nature that makes us what we are. We are the result of what we have decided for ourselves. Our evolution has led to pregnancy lasting 9 months, growth to adulthood lasting 15 -20 years, our chosen number of children to be no more than five and our usefulness after the children have grown up to be no more than a further 20 years, so we have almost decided for ourselves that 70 years is an acceptable lifespan. Once our children have been produced with our enhanced dominant genes, we have no purpose in survival of the species. We have produced a more suitable version of humans so we allow ourselves to die.

Over many generations this figure has become almost like an instinct and our genes and Telomeres have adjusted to phase out their activity to that lifespan. Perhaps we can change the lifespan simply by changing our expectation? If we expect to live to 100 and we can demonstrate our usefulness in doing so, then over many generations perhaps the lifespan will lengthen. Or perhaps we can do it by hypnosis and change the subconscious code that sets the length of the Telomeres.

Insects live only days but elephants live many years even though the construction of genes for both are similar, so we cannot say that our lifespan is fixed because all genes or cells gradually go wrong, instead it is the one central controlling gene that we have 'programmed' ourselves that decides to phase out cell division and life. It is this gene that causes brain neurons to trigger cell division and there is probably a fixed rate at which it does this. But this rate is programmed to slow as we get older.

In lifestyle terms the more new challenges we take on with enthusiasm, the more new exciting places we visit, the more we are motivated by a form of discontent with what we have, then it becomes more likely that we will be driven to live longer to put matters right.

But looking at in strictly biological terms, it means we need to find the central gene that controls cell division and record its energy output onto a memory chip. Then connect the chip so that it produces the signal to all parts of the body forever. This will not be achieved in my lifetime so I will settle for a contented lifestyle in the Bahamas!

I am not totally convinced about Telemeres (also known as 'tags'). The real cause must be somewhere in the process of energy diversion at puberty from growth toward sperm generation so that the signal code in the nerve that sets the rate of growth energy into a cell can only be produced sufficiently frequently for repair, and so growth stops. The reason is possibly that we have too many genes and copying all of them to make sperm exhausts our electrical energy and so new growth becomes impossible.

I read in February 2005, about some research done at Edinburgh and Glasgow that showed that people who have a fast reaction time live longer and this factor is even more important than IQ. Reaction time was the time taken to press a button after seeing a number on a screen. That suggests that it is something to do with motivation.

My conclusion is that death is what we want! It is due to a decline in motivation. It is the feeling that we have done it all before. We do not want more children. We do not feel the need to learn anything more. We are comfortable with our lifestyle and so the brain genes and cells that generate the growth somewhere deep in the subconscious slowly stop.

This not a question of lack of excitement as that is just a short burst of adrenalin, rather it is thinking that "My life is complete. I have done all I wanted to do". We are content and as I have suggested before it is discontent that drives everything in all animals, and as all animals strive for contentment, all animals will die.

So the problem is the exact opposite of what we think it is! We think the retirement dream is a beach-front house in a warm climate and when we have it we will be content for the rest of our lives. That is true but it is the fulfilment of the dream that contributes to ageing and death.

One can summarise all this and say we have decided for ourselves how long we should live. Evolution has demonstrated that sperm production is more useful in prolonging the species than new growth or a longer life, and if it were not so the change in process and gene would not have survived.

So can this instinctive process recorded in a gene be reversed?
I would prefer not to comment on whether life can be extended because
the planet cannot support the increase in population that would occur if
we all lived longer.

# Logic and creativity.

The left side of the brain is used for control and logic, and the right side is for creativity. The right side carries the detailed information required for such things as mathematics, music and art but the left side controls the level of detail that the brain will use to compute its decisions and outputs.

Scientists at Columbia University have shown that if magnetic pulses are focused on the left hemisphere to temporarily stop its function, 'control of detail' is removed and the person will immediately become more creative and mathematical. A person's personality can be changed temporarily.

In my case my left hemisphere is clearly in control. I am totally logical and whilst I play the piano and paint in oils fairly well. I have zero creativity. Everything I do in these fields is just a computer copy of what others have done before. The same brain functions were applied to reach the conclusions of my first book in which, using my pure logical, controlled and un-detailed mind, I questioned whether the detailed and 'uncontrolled' theories of right sided brain mathematicians could possibly be correct. To my brain their theories are in fairyland but who is to say who is right?

# Global Warming.

Planet Earth has gone through many changes in its four billion year lifetime but these changes occurred over millions of years and such changes will continue. The immediate problem is that man has multiplied in numbers and his use of energy has increased to a level that is causing a very rapid change to the gases in the atmosphere and consequently the temperature at ground level.

The problem will be self-correcting as are most things natural. Fish in the sea will die, land animals will die and finally man will be eliminated. The planet will have returned to where it was perhaps three billion years ago.

Stromatolites will restart to produce oxygen in the sea, then plants will grow, then animals and ultimately a new version of man. But that will take about two billion years.

The point is that man will never destroy the planet, he can only destroy himself. The planet will recover.

But if the result of man doing nothing to change his lifestyle is the extinction of the human race, then the most sensible course of action is to deliberately reduce the population to a level that allows the planet to continue unchanged. I.e. it is better to reduce births and live comfortably than to go through a slow and painful death until we are all extinct.

Reducing harmful emissions by 50% has little benefit if at the same time we allow the population to increase by 50%. The net gain is nil, and we should recognise that the goal for emission reduction may not be met because the standard of living will be reduced and self-interest will tend to take precedence over 'possible long term damage to the planet'.

Thus population control is the kindest and fastest way of saving ourselves from extinction. And this must be achieved, not by laws, but by mutual agreement of the majority. It must become regarded as a disgrace to produce more than two children.

## Biochemistry simplified?

Biochemistry is an extremely complex subject, but just glancing through a few books on the subject, I think it may have been made much more difficult than it needs to be. Engineering uses simple words of no more that six letters to describe the working parts of a machine, such as pistons, valves, gears, shafts, pulleys etc. Biochemistry uses similar short words for the main parts of the body, but a majority of very long words at cell level. When I searched motor neurons on the web, the first twenty lines included the following words.

Somatic, Parasympathetic, Acetylecholine, Skeleton, Intercostal, visceral, ganglia, monosynaptic, disynaptic, cholinergic, noradrenergic, etc!

I cannot even pronounce them let alone spell them or remember them. How can anyone pass an exam when it must use up a million brain cells just remembering how to spell them. Is it a device to build a barrier of mystique between doctor and patient? If so they should stop and instead use their brain cells to understand physics and chemistry rather better than they do now.

There is doubtless a valid source and legitimate reason for every such word but perhaps abbreviations are possible, or words that carry a direct and obvious meaning?

Mitochondria is the jelly and genes that surround the nucleus of a cell and provides the energy for a cell. So why not just call it the cell's power module? (Or in USA style, 'Power Pak'). Any engineer who had to call the cylinder that goes up and down in an engine as a pistochondrianol would pretty soon change it to piston. Keep it simple!

132

# Chapter Thirteen

## THE GOLDILOCKS RIDDLE

The Goldilocks riddle is; considering all the possible variations of processes and the laws of physics, how is it possible that exactly the right laws exist for life to have been created, even to the extent that the life that emerged is capable of having sufficient intelligence to ask such a question?

The smallest change in the size of a proton is said to make life impossible. I gave a brief answer to this in my first book but I will attempt a more thorough answer here.

People who seek a 'reason' for the universe and everything in it compound the issue. This is just a human emotion and this emotion is one factor that has led to the growth in religion.

In fact the universe can exist quite happily without us humans as I am sure many other planets show, and so no 'reason' is required. Physics, chemistry and biology have laws that exist without a need for a reason.

Equally there is no reason to puzzle how it is possible for a universe to create intelligent life that is able to think about how the universe came to exist. The creation of intelligent life is just automatic evolution as I have attempted to show in this book, so if scientists want to know 'why there is intelligent life' I hope I have helped a little.

My logic suggests that there is a natural law of creation. It is,
***Nature will always do the simplest and smallest thing necessary to do the job, and the job is to achieve equilibrium of energy everywhere'.***

An example of this is simply that hydrogen and helium – the two simplest atoms – were created first.

The reason is that every process in nature must be one of seeking equilibrium of forces and that means every step taken in the building of a universe must always take the simplest approach first. It is not possible for nature to jump straight into a complex structure because things evolve by using the simplest structure first to achieve the objective and then use more complex things only if that becomes necessary.

Of course just as the universe does not have a 'reason', neither does it have an 'objective' but there is an end point that its forces are trying to achieve and that is stable equilibrium. When that is achieved the forces and energy have nothing more to do.

## The multiple universe answer.

In terms of the universe, let me first dispense with one argument; the multiple universe theory. This suggests that there are many universes and each one has a different set of laws of physics. By chance our universe has just the right laws to enable life, but there are many other universes that do not.

My simple answer to this was contained in my first book and it is that there cannot possibly be any other universes. Beyond our universe there is nothing. No space. No dimension. If you are able to stretch your mind just a little then you can realise that such a place does not even exist.

In that situation two universes would touch and form one universe. And this would happen at every Big Bang when the universes were just energy. The energies would combine because there is nothing to keep them apart and there would be one Bigger Bang. Read my book if you would like further clarification.

## My theory of the universe

(Sorry but the following is physics, not biochemistry)

There can only be one overall structure for an atom. There can only be one hundred or so types of atoms, or types of material and the Earth has all of them. So the elements necessary for life would seem to be fairly easy to produce during the evolution of a universe after a Big Bang in which there are billions of galaxies and trillions of stars and planets.

In any universe there will always be stars large enough to achieve the density necessary for nuclear fusion and the creation of the many types of atoms.

In any universe the speed of light, the strength of gravity, inertia and mass will all be the same because these depend on the rate at which a charged particle produces a field and that is not a variable. All particles of all sizes produce a field at the same rate. Bigger particles produce more fields, but the rate of production does not change.

The fundamental question is what sets the size of a proton that is in the nucleus of every atom and radiates the electromagnetic field that is space and gravity, and by size I mean the charge of the proton?

The conclusion I reach is that the quark must be the smallest device that can form a store of energy and these stores became necessary because there was no spatial dimension (space) at the time of the Big bang.

The quarks are attracted to each other because energy had to produce an equal number of positives and negatives.

But two quarks of opposite charge could not achieve the objective of 'space creation' necessary for their existence. There has to be three so that a field of a specific polarity can be produced, and the field is the space dimension. It was the pressure of zero space that forced these three quarks together with 'binding energy' and prevented them annihilating themselves or pushing apart.

And because it takes three quarks to make space, (i.e. an overall positive structure) there has to be loose negative electrons to balance the charge to zero. The quarks could have formed into an overall negative structure in which case the electrons would be changed to positrons, but the effect on the universe and life would be unchanged.

Confirming my principle that the smallest and simplest structure are followed,

1. The quark is the smallest container of energy. (As far as we know).
2. Three quarks form the smallest device able to produce space. (The proton)
3. The proton is therefore the smallest basis possible for atoms.
4. The electron is the smallest positively charged particle that balances the negatively charged proton
5. The smallest atoms were produced first. (Hydrogen and helium).
6. The force of gravity cannot vary and so the density necessary to produce the hundred other atoms by nuclear fusion was inevitable.

It goes without saying, that if a quark had to be bigger to store energy it would simply make the proton bigger, and that would make no difference to the outcome because it is the rate of producing a field that sets all the laws of physics and that cannot vary, whatever the size of the particle.

Life was then just a matter of a planet and a sun being the right size, the right proximity, the right chemicals and water and life becomes automatic. No magic or intelligent creator is required.

# Chapter Fourteen

## SO - WHAT'S IT ALL ABOUT?

I did not deliberately set out to write two books back in 2003. My first book, *'A New Theory of the Universe'* was the result of reading Stephen Hawking's book *'The Universe in a Nutshell'* and finding that the theories of scientists that were reported in the book at that time seemed to me to be illogical. During the writing of my first book other rambling ideas on the evolution of life came to mind and it seemed appropriate to type them into a blog. That blog became the second book.

As I delved deeper into the scientific facts and the gaps in scientific knowledge, my quest turned into finding out how I believed things got to be the way they are.

Nature is completely dependent on the universe. Both use the same energies and forces that we call 'physics'. The evolution of life came about because of physics. The universe and nature have just one objective: To achieve equilibrium after the Big Bang.

But this objective is impossible to achieve. The larger the universe becomes the more complex the life forms within it, the harder it becomes to achieve equilibrium.

So in the end, billions of years from now, when nuclear fission in the stars has destroyed all the hydrogen and all the protons have exhausted all their energy, space, time, light and gravity will disappear, and with no space the universe will collapse to a tiny single point and a new Big Bang will start all over again. Galaxies will be rebuilt and evolution will start all over again.

And so what is the purpose of life?

There is no purpose, it is simply an automatic response to the presence of energy and chemicals. However one could summarise and say that if there is to be a purpose to life then it is not, as one might expect, 'to be enjoyed', it is 'to improve the human race' or 'to improve survival' and that is what each of us should strive to do.

## The Big Questions.

The question is did all this come about due to some weird external force or intelligent design or was it all automatic and inevitable?

Most people's answers will depend on what they would like to believe, (i.e. an emotional response) or what they were taught to believe, but I do not fall into either category. I simply decide from my own unemotional logic and analysis of the facts, which cause is the most likely.

Why is there a need for a reason for man to exist? It is an emotional question, not a logical one. The answer is the same as the number '16'!

The number 16 exists because there are smaller numbers, and when combinations of these small numbers are added together they reach number 16. So it is with man. There are smaller chemicals and life forms and when added together, man is produced.

And just as number 16 has evolved and produced 17, 18 and 19, then so man will evolve and produce 'better man', 'outstanding man', and 'super man'.

In fact I think that intense competition in future will generate the need for specialisation for survival, so that whole generations in a family will have the same profession, such as sport, politics, languages, science, and maths. Without such specialisation evolution will not provide a child with the necessary genes needed to succeed. And, instead of streaming children in schools based on academic skill, they will be streamed on intended profession, as chosen through the family history.

My conclusion after examining the universe in my first book was that it was an inevitable outcome from energy. No other factors were involved.

My conclusion from the analysis of life forms in this book is the same. 'gene adjustment' and 'use' played a role in causing the evolution from a basic life form to an intelligent human.

The process was very similar to that of the design of the motorcar. If we were super-intelligent we would have produced the perfect car in the early 20th century, but instead we have gradually improved the design by learning from mistakes and trying new things.

The same argument can be applied to all forms of life. It is easy to see how each life form produced improvements on earlier models to arrive at where they all are today, and there is no reason to think that the process is not continuing.

One could argue that 'intelligence' of a kind was involved but it was not a grand overseeing external intelligence, it was the intelligence brought about by the ability to choose made individually by each life form, led by the desire to seek food, protect itself from predators and survive.

The driving force for evolution in animals was a matter of seeking constant 'contentment'. The opposite of contentment is fear, hunger and pain, or in short, every adverse thing that the senses told the brain, no matter how small the brain.

This contentment obviously does not arise in the plant world but then I would argue that the existence of a plant is purely a chemical reaction to available energy. Mutations provided the ability to adjust and survive.

# So where does all this leave religion?

If the Big Bang was inevitable and the evolution of life was inevitable, where does that leave religion?

My conclusion that the universe will collapse automatically, and then expand in a never ending sequence of Big Bangs, indicates to me that there is no requirement for a God or Allah. Why would an intelligent force such a God destroy a creation with all its varied life forms, and then re-create it all over again?

So it is absolutely clear to me that there is no God or Allah, and there is no 'afterlife' of Heaven, Paradise or Hell. There is only this life that we have now.

If a God or Allah was not required for the creation of anything, what is the purpose of religion in society?

I believe the background that led to religion can be split into two parts,
1.  The belief that God created the universe and all life within it.
2.  The teachings of religious leaders in how to be a civilised person.

These two parts became mixed together inevitably by the religious leaders in order to add some meaning and enforcement to their teachings in how to behave. It adds opportunities and risks to right and wrong. But we have to remember that this concept grew in the time when there were few laws, no police and no schools.

Now that schools, laws and police exist it is possible for some of the teachings of good behaviour to be implemented without the need of the threat of God, Heaven and Hell to enforce them.

I suspect that schools prefer to retain the concepts of a 'creator' and 'intelligent design' because without them the link is broken and the Bible becomes meaningless. The foundation of right and wrong and the feelings of conscience and guilt disappear. It is much easier to quote from the Bible.

But the problems of religion that exist today are largely because the 'teachings' are different for each religion. There is no common or agreed standard of behaviour and lifestyle. Some are strict disciplines while others allow considerable freedom to act and rely on laws for control, ultimately leading to the 'nanny state'.

# The future of religion.

It is interesting to note that scientists have discovered recently that a belief that a medicine will work actually triggers the release of painkillers in the brain and this helps to explain the placebo effect. When patients expect a treatment to be effective the part of the brain that is responsible for pain control is activated. It follows that if you really believe that a person such as Jesus, has the power of healing, this simple belief is sufficient to produce favourable results. Faith is a very strong tool.

One can even extend this concept further and say anyone who has a positive frame of mind and is content with life is less likely to be ill.

If a person needs to attend a church in order to retain faith and keep this 'placebo effect' going that is fine, but the error comes when one draws the conclusion that the placebo effect is due to an external force, or God, when as scientists have just shown, it is due entirely to the brain, or inner self.

Man has always needed spiritual help and has tried 'Sun Gods', stars, pyramids, voodoo, witch doctors etc. and most are driven by attempts to survive famine, illness and poor crop yields. There will always be a role for religion because there will always be a need for hope, love and friendship, but it should not be something you have because you are told to; it should be something you feel that you need. It is an emotional requirement.

In my atheistic case, when things are bad and I need spiritual help I walk through a pine forest high on a hill and just sit and admire nature while smoking a cigar. When I walk back down I am relaxed and refreshed and whatever the problem was, it has either gone away, solved, or no longer seems important. That is exactly what both Jesus and Muhammad did, except without the cigar! And the so called 'revelations' of Muhammad are what I would call 'brainwaves'. I got them regularly while trying to solve the universe, and they usually happened at about 5 am. It is amazing how the brain can solve a problem while you are asleep but it has nothing to do with God.

I am 100% in favour of churches as excellent places for peace, friendship and social gatherings, and preachers who teach right from wrong. I can fully understand that it was necessary to emphasise good and bad to young simple minds by introducing the concepts of heaven, hell, God and the Devil. But these concepts are just 'tools' to assist teaching and the concepts themselves are rather like Santa Claus and fairies. Whether you believe them or not depends on your particular circumstances.

According to Dr. Robert Beckford, who has carried out substantial research of different religions, the name of Jesus occurs in most religions but each religion regarded him differently, i.e. some as the Son of God, some as an ordinary man who was a good teacher, etc, so we do not really know who Jesus actually was.

Also one has to remember that Rome was very selective about which writings should be included in the New Testament of the Bible, and which should be deliberately excluded because they did not convey the story that Rome wished to present – i.e. that Jesus was the Son of God. Because of such editing one cannot be certain whether the writings that are included are exactly accurate, or are simply chosen interpretations of what four people think happened, whilst the unfavourable differing opinions of another ten people were excluded.

I have always had doubts about the nativity play that school children perform at Christmas. Exactly who was the Angel Gabriel? If it was so easy for him to visit at that time why doesn't he come back now and clarify things? How is it that the three kings had gifts of gold, incense and myrrh readily available on their donkeys? Were they expecting to find someone they could claim to be the Son of God? Was the Angel Gabriel just an actor being paid by the kings to confirm that their prophecy of the coming of God had come true? The Archbishop of Canterbury, the head of the English church, said in 2007 that the nativity is all just legend. There were no 'three wise men'. Nothing in the bible suggests that there were, or that there were oxen and asses in the stable. So are we pretending to our children that the Nativity is true, when much of it is legend?

But this was the single act that started the whole of Christianity as it was not Jesus who claimed to be the son of God, it was the Angel Gabriel. If one doubts the existence of the Angel and questions the accuracy of the bible, then belief becomes solely a personal matter, influenced by circumstances.

Clearly those people who live a life of poverty, illness or extreme unhappiness need hope and love and these people are motivated to believe that an external force can change their desperate situation. It is the reason we all do the lottery. But this does not mean that there was some kind of 'master plan' that caused things to be the way they are, or that there was an external force that made it all happen.

The concept is the same as 'superstition'. The belief that soccer players have a lucky shirt or lucky boots that if they wear them they always seem to result in more goals. And these concepts seem to be most common in the hotter regions of the world, either because the sun causes particular emotions to become strong or because the lack of rain causes the very hardships of draught and disease that require hope for life to be worthwhile.

The problem for the atheists who seek to convince believers that there cannot possibly be a God is that the belief has lasted for so many generations that it has become an instinct in the genes. It is like trying to convince a bird that it does not need to migrate. There is a book by David Brooks entitled 'The Social Animal' that argues that man is driven more by instinct than by rational thought.

## The problems of religion.

The problems arise when rivalry occurs between religions that have slightly different beliefs or standards of good behaviour. The West believes in freedom but the converse is that this very freedom allows people to choose to do things that others would regard as wrong. The East believes in strict behaviour with very little freedom but the converse of this is the lack of independence to decide for oneself.

This extreme discipline is the very opposite of the West causing westerners to think that kind of religion is not in the best interest of the population and is wrong. But the West freely allows porn on TV and hundreds of porn websites and so it is no wonder that the East despises the Western 'freedom' and their importance placed on wealth and material possessions.

Another problem with the Bible is that it can never keep up to date with events. It does not provide any guidance on whether it is right to jump a red light, exceed a speed limit, use a mobile / cell phone whilst driving, take drugs or copy DVDs. Does that mean doing these things is OK?

## Religion thwarts progress.

The key point is that if 'contentment' is achieved from belief in God, then there is no incentive to develop new products and improve ones material life. But the most serious problem with religion in my opinion is that it is 'brainwashing' and it prevents the progress of a civilisation. Evolution has given man the ability to think and the worst thing a parent can do is to remove that freedom to think in order to teach conformity to their own beliefs. It will produce a well ordered and disciplined society but the individuals within it are little better than the robotic ants I described at the beginning of the book. And instead of seriously thinking about how to solve a problem by oneself, religion leads to time spent hoping for heavenly intervention. Thus religious countries are slow to develop. They also have very little need to because they gain contentment through worship and friendship rather than technical innovation. Indoctrination thwarts progress. But worse than that, if religion is brainwashing it is dangerous to allow any religion to form into a political party and to lead a country.

But the converse of this is the possibility of the downfall of less religious societies where competition, jealousy, money and bribery flourish. Freedom leads to crime and ever more Big Brother state intervention.

We need something in between. Not Big Brother. Not a dictatorship. Not a bureaucracy.

But if we can get back to the basics and understand that we are all individuals that have been given the ability to think for ourselves. If we can accept that there is no God and that the early teachers were simply setting out guidelines for behaviour that they felt would bring the greatest level of happiness to everyone, then we should accept that they were just guidelines by intelligent good-meaning people and not rules laid down by God.

By all means read the early teachings for inspiration but surely it is right for each person to think and decide what is right for them in today's environment with no brainwashing or pressure from others. Then it does not matter whether the beliefs are different; what matters is that each person was free to make that decision without any forceful threats to obey some other doctrine.

We should respect freedom and respect those people that have decided to adopt a nice way of life. We should penalise those people who seek to enforce their beliefs, including racial hatred, on others and those who have chosen to use their freedom to make life unpleasant for others.

One-way to achieve this is law after law but so many laws would be required that it would become unenforceable. A better way is education and early discipline to guide young malleable minds in the proper way to behave to produce the maximum happiness and prosperity for everyone. Is that the role of parents, the school or the church?

In our western society more and more people seem to cheat or break the law, often at the expense of others. This seems to be driven by competition and leads to jealousy and envy. But there can never be enough police. It would take 50% of the population to police the other 50% and it would be state control like Big Brother. Yet if you stop celebrating victory, competition is removed and there is less incentive to improve anything, society would stagnate.

## The solution?

Firstly, the police are inevitably limited in numbers and their role should not be so much to prevent crime, but rather to ensure that anyone who breaks the law is always caught. It is the role of parents and the community, and to a lesser extent schools, to educate right and wrong and in so doing prevent crime, and civilisation depends on this role being carried out effectively. Animals have always been doing it (without their own police force!) and parents might perhaps re-learn from them, how to do it. A gentle nip from a kitten's mum seems to be their solution to keeping order.

Secondly, the solution would seem to be in 'specialisation'. Everybody can excel in one thing and have pride in it. So the key is an education that enables every child to try everything until they find out how they excel.

But there needs to be a fundamental change in religion in schools. There should not be any 'faith schools'. The role of Jesus should be placed within the history syllabus and the syllabus of 'Religious Instruction' should be changed to 'Behavioural Instruction' where all the good teachings of Jesus, Moses and Muhammad are taken out of their religious context and taught as moral guidance and behaviour for a civilised community.

This may take years, but the end result will happen eventually whatever our leaders do because the increasing intelligence of the population will ensure that it is so. The greatest force in bringing about improvements in the relations between people is the community itself.

# And finally

In this book I have suggested that contentment, or rather discontentment, is the force that drives animal life forms to evolve. A cow will always be a cow because it is content. It receives all that it requires in food and protection.

But what about 'man'? Have we designed the ultimate person? We too have adequate food and protection, so has our evolution stopped?

My suggestion is 'no', but the future depends on what we want. Do we want to run faster and hit a golf ball further, or just sit and be carried everywhere by car? Do we want to communicate with relations far away by telepathy instead of by mobile / cell phone? The choice is ours!

Man is an intelligent, adventurous and competitive animal. We know better things are possible and so we will never be content.

*The evolution of man will continue*.

# And a final twist.

My book has reached the inevitable conclusion that mankind is just a further step in the evolution of animals, but that is not really what I want!

I have no wish to have DNA in common with a daffodil, or to realise that I am just a slightly more intelligent ape. I want to feel special and completely unrelated to pigs, cows, sheep and fish. I like them all but I do not want any direct links as a descendant with them! An education and smart clothes begins to make me feel remote from the ways of animals, but it is not enough.

So perhaps I am getting to the real point of religion that wise atheists such as Richard Dawkins have not yet grasped. If people wish to choose a path in their lives that has no relationship with animals, then they must believe that mankind was created by a super power – A God.

To be 'special' requires belief and disciplines that people choose to follow, not because they are right, but because they make them unique and remote from the cruel, sordid and muddy animal world.

But for me, I am happy to just draw a line at year zero AD and believe that that is when man learned the discipline necessary to control his animal instincts and ceased to have any connection with past life-forms. It was a rebirth of the mind rather than the body.

And that suits me fine! I am no animal! Well done Mr. Jesus, Mr Moses and Mr Muhammad.

Please tell your friends that as a result of reading this book your IQ has gone up at least 20 points and your spermatozoa is really excited about doing the job nature intended. And if you have a major problem that needs a solution, just fly to Exuma, Bahamas, stay at The Augusta Bay hotel and sit under a palm tree with a big cigar and the solution will come. Alternatively, find me, and for $1m I will find the solution for you while you relax on the beach.

My thanks go to the Bahamians for allowing me to live in their beautiful country, to Imperial College for teaching me how to think, the makers of big cigars, The Times newspaper for its regular articles on health, and my wife whose death from bowel cancer led me to ask why?

...................................................................

If you enjoyed this book may I suggest that you read my first book *'A New Theory of the Universe',* which I think is a better book!
The book explains gravity, space, time, dark energy, dark matter, black holes and many other aspects of the universe. It is available from the publisher – lulu.com - or most on-line book sellers now.